W0246776

Developing Microservices Architecture on Microsoft Azure with Open Source Technologies

Ovais Mehboob Ahmed Khan
Arvind Chandaka

Developing Microservices Architecture on Microsoft Azure with Open Source Technologies

Published with the authorization of Microsoft Corporation by:
Pearson Education, Inc.

ISBN-13: 978-0-13-681938-7
ISBN-10: 0-13-681938-9

Library of Congress Control Number: 2021937949

1 2021

TRADEMARKS

WARNING AND DISCLAIMER

SPECIAL SALES

For information about buying this title in bulk quantities, or for special sales opportunities (which may include electronic versions; custom cover designs; and content particular to your business, training goals, marketing focus, or branding interests), please contact our corporate sales department at corpsales@pearsoned.com or (800) 382-3419.

For government sales inquiries, please contact governmentsales@pearsoned.com.

For questions about sales outside the U.S., please contact intlcs@pearson.com.

CREDITS

EDITOR-IN-CHIEF
Brett Bartow

EXECUTIVE EDITOR
Loretta Yates

DEVELOPMENT EDITOR
Rick Kughen

SPONSORING EDITOR
Charvi Arora

MANAGING EDITOR
Sandra Schroeder

SENIOR PROJECT EDITOR
Tracey Croom

COPY EDITOR
Rick Kughen

INDEXER
Cheryl Ann Lenser

PROOFREADER
Abigail Manheim

TECHNICAL EDITORS
Doug Holland
Thomas Palathra

EDITORIAL ASSISTANT
Cindy Teeters

COVER DESIGNER
Twist Creative, Seattle

COMPOSITOR
codeMantra

I would like to thank my family for supporting and encouraging me in every goal of my life.

—Ovais Mehboob Ahmed Khan

I dedicate this book to my family and friends for their everlasting encouragement and support.

—Arvind Chandaka

Contents

Acknowledgments

This book involved great effort on our part. With the recent growth in microservices architecture, we saw the importance of developing a sample case study on microservices to empower our customers to build better applications. We created a use case to illustrate microservices, created a functional application, and evangelized it to our customers—all before even putting pen to paper to write this book. Our friends and family were the spectators during these laborious endeavors, and they showed nothing but encouragement and understanding during long nights and tiring weekends. We couldn't have done it without them!

A big thanks goes to our managers for their everlasting support. Barbaros Gunay, Andrew McCreary, Mekonnen Kassa—thank you! Without your support, this book could not have been done.

When we both started working at Microsoft, we were quickly taken under the OSS Enablement Team's wing. Rick Hines and Bahram Rushenas are the team leads who encouraged us to pursue the project that ultimately became this book. Whether it was discussing a design or architecture of an application, brainstorming ideas, or presenting at events, their help and support were tremendous. We want to thank you for everything you've done.

We also want to thank Sally Brennan for preparing our session outline for Microsoft Ready, our premier go-to-market conference, which led us to further refine the content we used within our book.

We also want to thank our team at Pearson. Thank you so much, Loretta Yates, Charvi Arora, Rick Kughen, Doug Holland, Thomas Palathra, Abigail Manheim, and Tracey Croom for supporting us during our journey. We would like to thank Rob Vettor for lending his time to review our book and providing a foreword. Rob is an extremely experienced individual in microservices at Microsoft and we greatly appreciate his support.

About the authors

Ovais Mehboob Ahmed Khan is a seasoned programmer and solutions architect with nearly 20 years of experience in software development, consultancy, and solution architecture. He has worked with various clients across the United States, Europe, Middle East, and Africa, and he currently works as a Senior Customer Engineer at Microsoft, based in Dubai. He specializes mainly in application development using .NET and other OSS technologies, Microsoft Azure, and DevOps.

He is a prolific writer and has published several books on enterprise application architecture, .NET Core, VS Code, and JavaScript, and he has written numerous technical articles on various sites. He likes to talk about technology and has delivered various technical sessions around the world.

Arvind Chandaka is a product manager at Microsoft and has led products across Azure and Cloud + AI Engineering. He has worked with strategy executive teams to grow Microsoft Azure as well as built IP at the company resulting in several services and products. He is a recognized SME in enterprise technology and has advised many Fortune 100 companies and international clients around the world. He specializes across infrastructure, identity, cyber security, open source, and more. A strong believer in empowerment, he works closely with startups, is a founder himself, invests in the entrepreneurial ecosystem, and serves on the board of non-profits in New York City.

Arvind earned his Bachelor of Science at Cornell University where he studied several disciplines ranging from computer science to business. In his spare time, you will likely find him at a tennis court or traveling the world for the next great foodie destination.

Foreword

For many years, the guidance for building business applications was clear-cut:

- Create a large, monolithic core that contains business logic
- Persist data to a shared relational database
- Expose functionality through front-end UIs and API endpoints
- Build, package, and deploy the application, and you're in business.

While often challenging, such monolithic applications were *straightforward* to build, test, deploy, and troubleshoot. With code executing in a single process, performance could be good. When the need arose, you could scale the application up (vertically) by adding more server resources.

This model, however, had limitations. When an application proved popular, the monolithic architecture would eventually break down. As the app grew larger, more complex, and more "coupled," it became less agile:

- Feature updates and fixes would cause unintended and costly side effects, such as breaking existing functionality.
- One unstable component could crash the entire system.
- Scaling any component required scaling the entire application.
- New technologies and frameworks were not an option.
- Each change required a full deployment of the application.

Perhaps the biggest challenge was *agility*. Monolithic applications prevented the business from moving fast.

Fast-forward into the world of modern apps and microservices. They are all about speed and agility. By decomposing the business domain into independent services, microservices enable business systems to strategically align with business capabilities. They embrace feature releases with high confidence. They allow for updating small areas of a live application without downtime. They empower the business to adapt rapidly to evolving market and competitive forces.

While microservice adoption is rapidly increasing, not everybody is adopting it correctly. Implementing a microservices architecture requires substantial investment. For it to be successful, developers and architects must understand the principles, patterns, and best practices of distributed microservices architecture.

This book *spoon-feeds* microservices to you. From design to development and from deployment to operation, it provides a comprehensive education into the world of microservices. Moreover, the book wraps microservice construction into the context of cloud native design and open-source technologies, all hosted in the Azure cloud. Kudos to authors Ovais Mehboob Ahmed Khan and Arvind Chandaka for making this reference available. Kudos to you, the reader, for investing the time to study and master the architecture.

—Robert Vettor
Principal Cloud Solution Architect
Microsoft

Introduction

Welcome to *Developing Microservices Architecture on Microsoft Azure with Open Source Technologies*. Today, organizations are modernizing application development by integrating open-source technologies into a holistic architecture for delivering high-quality workloads to the cloud. This book is a complete, step-by-step guide to building an application based on microservices architecture by leveraging services provided by the Microsoft Azure cloud platform and a litany of open-source technologies such as Java, Node.js, .NET Core, and Angular.

This will be a reference guide to learn key building blocks of microservices architecture by representing a real-life case study of building a trading system to auction items. This book will lead readers through a step-by-step journey to learn modern application development scenarios and leverage cloud-native capabilities in Microsoft Azure to provide distinct value to customers.

This topic is in very high demand based on statistics and our own experiences from the field. There have been books made on microservices, but they almost always cover business value proposition alone rather than technical implementation on how to establish end to end infrastructure, application development, deployment automation and to realize actual value through constructing it using Azure and open source.

We start with an introduction and modeling of microservices and then move on to build a cloud-native microservices architecture by developing services, containerizing services, and deploying them on AKS. We also cover various topics related to communication patterns, security, and API gateways, and we touch on topics related to automating builds and deployments using CI/CD and monitoring.

Who is this book for?

This book is for architects and developers. It especially targets those who have some development background to build applications or have worked in the capacity of designing and architecting applications. It will also be useful for those with some skills with Azure.

How is this book organized?

This book is organized into three parts:

- Part I: Introduction and modeling of microservices
- Part II: Designing and building microservices architecture
- Part III: Implementing patterns, security, DevOps, and monitoring

The first part of this book consists of two chapters that focus on a thorough introduction to microservices and how you can decompose or model microservices using domain-driven design (DDD).

The second part of this book consists of three chapters that focus on designing an architecture and building an application that follows this architecture with various open-source technologies such as Java, .NET Core, Node.js, and Angular. The architecture is cloud-native and leverages multiple managed services within Azure.

The third part of this book consists of five chapters that focuses on topics related to communication patterns, security within microservices, API gateways, automating builds and deployments, and monitoring.

System requirements

The requirements for running through the examples in this book should be:

- Any Linux distros or Windows operating systems
- 4 GB of RAM
- 1 GB of available disk space
- Docker with Linux Containers
- Microsoft Azure subscription
- Azure DevOps account
- Visual Studio Code or other editors of your choice

About the companion content

The companion content for this book can be downloaded from the following page: *MicrosoftPressStore.com/DevMicroservices/downloads*

Errata, updates, & book support

We've made every effort to ensure the accuracy of this book and its companion content. You can access updates to this book—in the form of a list of submitted errata and their related corrections—at:

MicrosoftPressStore.com/DevMicroservices/errata

If you discover an error that is not already listed, please submit it to us at the same page.

For additional book support and information, please visit *MicrosoftPressStore.com/Support*.

Please note that product support for Microsoft software and hardware is not offered through the previous addresses. For help with Microsoft software or hardware, go to http://support.microsoft.com.

Stay in touch

Let's keep the conversation going! We're on Twitter: *http://twitter.com/MicrosoftPress*.

Chapter 1

Introduction to microservices

In this chapter, you will:

- Learn about the history and evolution of software architecture
- Understand the differences between monolithic and microservices architectures
- Learn about the core benefits and challenges of using a microservices architecture

Microservices is a buzzword that is thrown around frequently in the technology community. Many infer it as being a further dissection of modular componentization, and although that addresses part of it, it doesn't encompass the full picture. Microservices are a group of back-end services that provide business operations to form an application. When combined with other pieces such as a front-end and various communication platforms, it creates what is known as a *microservices architecture*. Although that is a simplistic way to approach this architecture, it is a vital design orientation and metaphor to learn for implementing modern application development scenarios geared to achieving better results.

This particular chapter will focus on the core fundamentals of microservices, including the transition from historical architecture to present, benefits and challenges, and a comparison of a monolithic application architecture with microservices architecture.

Our journey with microservices

In recent years, microservices architecture have become very popular with firms. Many organizations and enterprises have expressed a desire and need to move their applications to a microservices architecture.

We have worked with customers that were very concerned with increasing development velocity to ship code faster and at more frequent intervals. Other customers are focused on modularizing their codebase further to iterate on particular services for better application performance and additional support. Some are jaded by financial and time constraints that constantly plague their organization due to large scale migrations or refactoring and want to put an end to it. We have seen these pains, frustrations, complex scenarios, and wishes from a myriad of customers time and time again.

We have worked at Microsoft and in the IT space with many enterprise customers. We met with various internal open-source groups to find solutions to these problems of customers. Microservices architecture was the key behind this, and ultimately, we wanted to have a stream-lined way to demonstrate the theoretical value of microservices and how to build one, too.

As a result, we created an application called the Online Auction System (OAS) to serve as a real-life case study to teach and empower customers to build this architecture for their enter-prise scenarios. We accomplished this by using a variety of open-source technologies, and we used Microsoft Azure as the bedrock for the application. Throughout this book, we will use the approach to provide this background knowledge in conjunction with the parallels to this case study, which helps map theoretical and experiential learning together.

Evolution of software architecture

Before we can understand the details behind microservices architecture, let's take a brief look at how software architectures have evolved.

Monolithic architecture

About a decade ago, the monolithic application was very popular, and consequently its N-tiered architecture backbone was a norm. This consisted of an application decomposed into several layers. These layers vary depending on the complexity of the application, but tradition-ally, there were three main components. See Figure 1-1.

Monolithic architecture

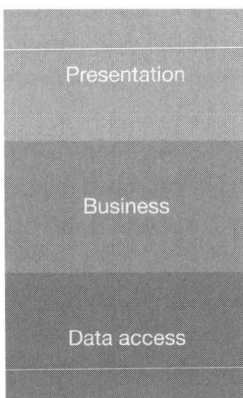

FIGURE 1-1   The monolithic architecture model

- **Presentation layer**   The presentation layer traditionally represents the front-end. Some of the artifacts include view models, view pages, user interfaces, and other artifacts that are oriented toward front-end development.

- **Business layer** The middle layer is a business layer that holds all the logical flows of the application. This includes most of what we consider as the back-end code necessary for the primary function of the overall application. This area is where languages such as .NET or even Java are used to build back-end services that contain business managers and business objects consisting of the application logic.

- **Data access layer** Finally, the data access layer is responsible for all database operations. This consists of a database and interface to perform read/write operations on data collected and used within the application as part of its overall functionality.

  This layered architecture provides a basic separation of concerns and decomposition of applications into functional components. However, this architecture results in tightly coupled services that result in many dependencies.

Thinking of some of the scenarios from earlier, we can see why tightly coupled services and dependencies cause problems. Dependencies present drawbacks where it is difficult to make any changes to the application easily. This includes complexity in the form of large migrations and refactoring projects where many teams are involved and many times agility is replaced with a potential for bureaucracy. This introduces longer timelines because simple quick deployments cannot be made and ultimately all these problems cascade into a single point where management is then required to make massive expenditures.

Service-oriented architecture (SOA)

The transition away from monolithic architecture was driven by these reasons. The heavyweight nature of this architecture was slowly decomposed and eventually led to the formation of a new architecture known as service-oriented architecture (SOA), which is illustrated in Figure 1-2.

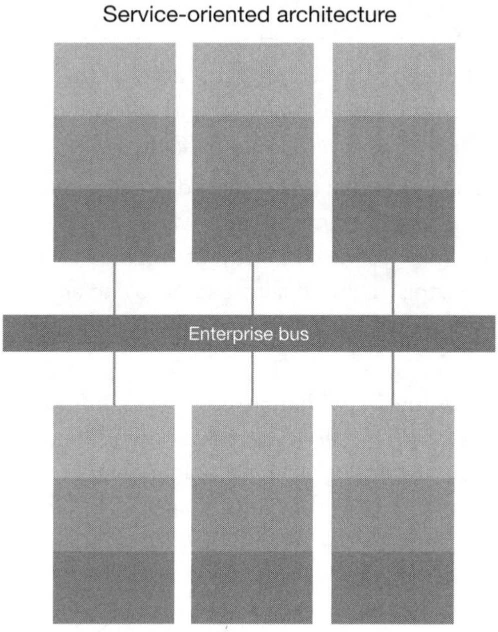

Service-oriented architecture

Enterprise bus

FIGURE 1-2 SOA architecture

This model was geared toward decomposing an application into even more granular modules to account for the drawbacks in monolithic architectures. Essentially, SOA was a way to connect different monoliths across a consistent messaging channel. Thus, as you can see in Figure 1-2, with SOA, you can have a set of services that are serving various functions connect and communicate with each other over a service bus. This service bus essentially is the communication tool that assists with message routing, transformations, security improvement, logging, and much more.

Because the services are broken down by function, it is slightly easier to deploy changes that don't affect the whole application. So, if you were building an application with a monolithic architecture, you could refactor it by using SOA architecture and addressing the litany of cascading problems we had with the monolithic architecture. Although SOA addresses some of these problems, we haven't solved a primary pain point in either architecture model—assuaging application scalability concerns.

Application scalability is an important design consideration when thinking about the growth of the application over time. With a larger user base, you will likely have more complex requirements, and thus considerations must be in place for handling additional requests and traffic. The clunky nature of the monolithic architecture doesn't address this and neither does SOA. Ultimately, with more scale, SOA breaks down internally because the single enterprise bus is overwhelmed, which slows (or throttles) during periods of increased requests and traffic.

That being said, the SOA concept ultimately introduced an important integration solution that helped applications talk to each other, and the additional consideration of scalability gave way to the expansion of the microservices architecture.

Comparing monolith with microservices

A microservices architecture is a series of services that communicate with one another to achieve complete application functionality. Each microservice has a similar pattern to a monolithic architecture but only covers one function within the entire application functionality.

This means that each service can have a data access layer that is customized to use several kinds of database technologies and a business layer that is customized to use wide varieties of technology to perform the functional aspects of the microservice. To elaborate on the business layer using our case study detailed in future chapters, this means that instead of having all the OAS back-end code in a single layer, we have individual services for setting up an auction or placing a bid.

The microservices architecture iterates on previous models with more communication and an all-encompassing front-end functionality. All the back-end services now communicate with each other to represent actions that users can experience through a single point of entry within the front-end, as illustrated in Figure 1-3.

Microservices architecture

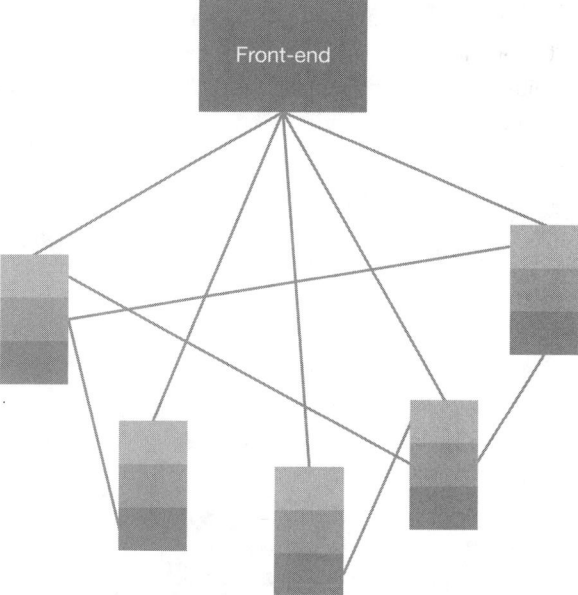

FIGURE 1-3 Microservices architecture

SOA versus microservices

You might also wonder how SOA differs from microservices. Microservices are fine-grained services where each service is modeled for a business domain or capability. SOA services are usually coarse-grained services that are not completely decomposed based on the business domain. Microservices are more abstracted and segmented by the individual business functionality. We will cover this in detail in future chapters, but the use of domain-driven design is a major design factor in microservices that is distinct from other architectures. Domain-driven design specifically aligns services to individual business functions in an application.

Also, microservices have more complex patterns when it comes to communication and messaging. Where we once had an enterprise bus as one of the innovative features of an SOA, we now have a broader set of technologies and mechanisms, such as request-response and publish-subscribe communications. These patterns will help distribute communication channels for better performance in an application, rather than making the service bus the bottleneck of the system.

Another advantage of microservices is rooted in using lightweight protocols, such as HTTP, whereas with SOA, we use other protocols such as TCP or MSMQ. There are many examples in which microservices are the presumptive choice for your application architectures. To help illustrate this better, we will dive deeper into some of the issues with legacy architectures.

Monolith example

Now that we have talked about the high-level differences, let's run through an exercise to understand the minute differentiations between monolithic and microservices architectures. To help illustrate this, see Figure 1-4, which illustrates how the monolithic architecture requires a full application deployed in each host, making it difficult to scale.

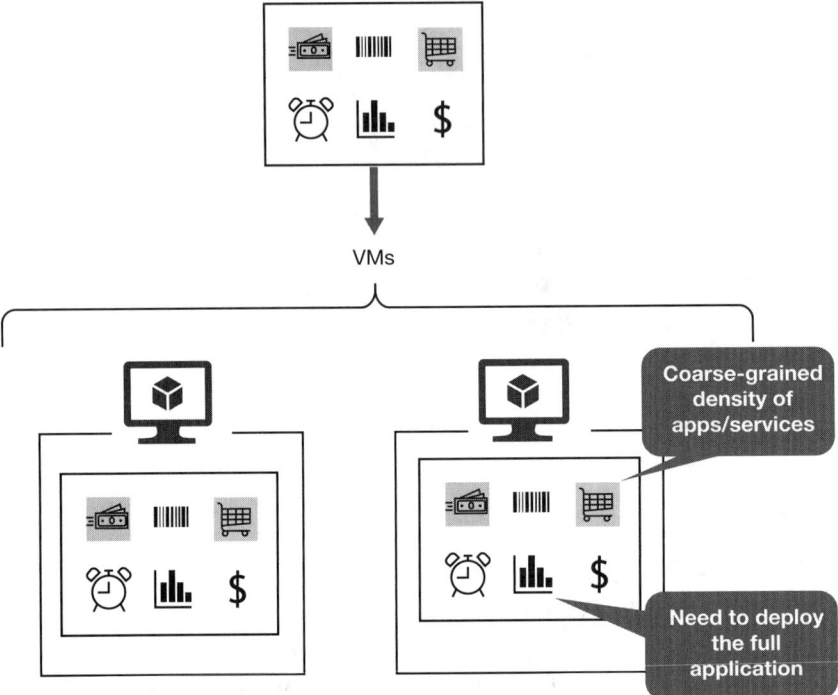

FIGURE 1-4 The monolithic architecture is difficult to scale

Figure 1-4 helps clarify the nuances of a monolithic application. Here, we have a collection of services that are ultimately deployed on two virtual machines (VMs)—all containing the same services within each host. As you can see, the scaling solution in place essentially deploys the full application on multiple VMs.

> **NOTE** We know the monolithic architecture is scalable, however, it is expensive. We need to have the prerequisite specifications for these VMs with the same memory, compute power, and so on. Even though the deployment of the monolithic architecture might be simple because it is consistent, this can be expensive, whether you are utilizing your on-premises datacenter for the infrastructure or you are using a cloud provider with a pay-as-you-go subscription.

Microservices example

Now let's compare the monolithic architecture with a microservices approach. Figure 1-5 illustrates how the microservices architecture can be deployed independently in several different hosts and is more capable of scaling.

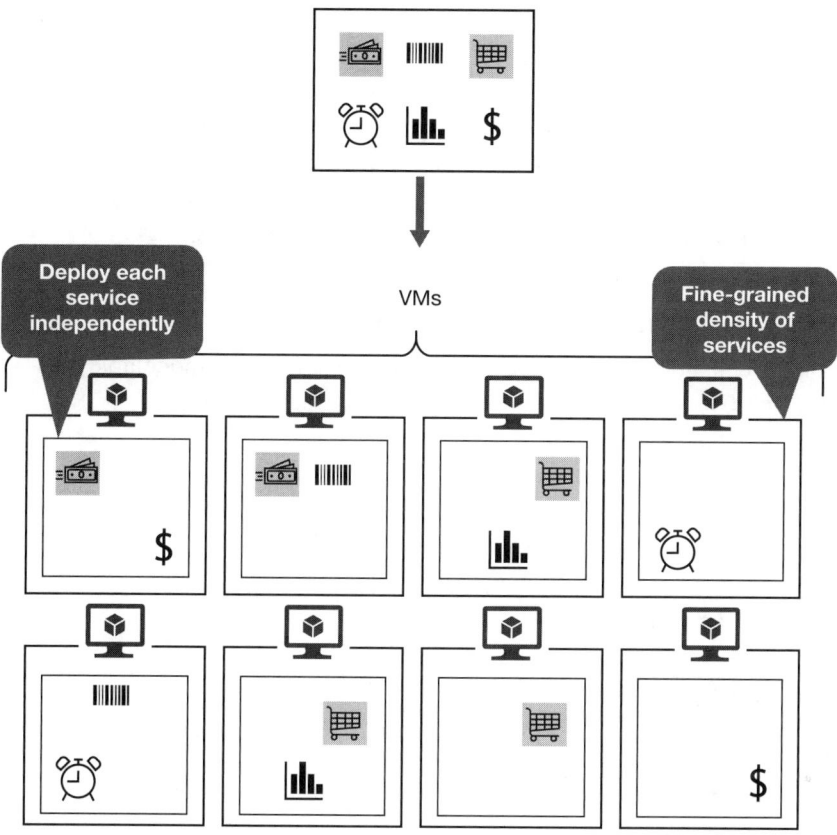

FIGURE 1-5   Microservices architecture

Figure 1-5 shows how microservices architecture differs from the monolithic approach. Virtual machines can be deployed in the microservices architecture, but instead of deploying all components of the application in a single VM, we see that each service is deployed independently on several VMs. Furthermore, with a monolithic application, any deployment that might have issues or bugs would have to be rolled back and analyzed carefully for mistakes. Not only is this tedious, but it provides a bad user experience for your app's consumers because of bugs and unexpected downtime. This allows us to scale based on functional need, doesn't result in overexpenditure of your resources, and enables a more fine-grained control of your services. If you experience deployment issues, you can easily find errors and redeploy with flexibility and velocity.

To understand how microservices promote scalability, suppose we are running an order management system in which users can place orders. On weekends, the system load is increased by a high number of user orders, which causes response timeouts and improper handling of resources. With the appropriate monitoring framework in place, we can identify when the order service is not handling a larger amount of requests. Based on this, we can just scale out the order service, rather than scaling out the whole system. This allows us to save our resources inexpensively.

> **NOTE** Although we are demonstrating some of these basic concepts with microservices spread across VMs, a more regimented enterprise approach will use infrastructure such as Kubernetes, containers, and continuous integration and continuous development (CI/CD) as components to bolster some of the microservices benefits.

Databases in a monolithic architecture

Let's focus more on the data tier of the architecture. Figure 1-6 shows that a monolithic database might not be a good design choice when it is shared across many services.

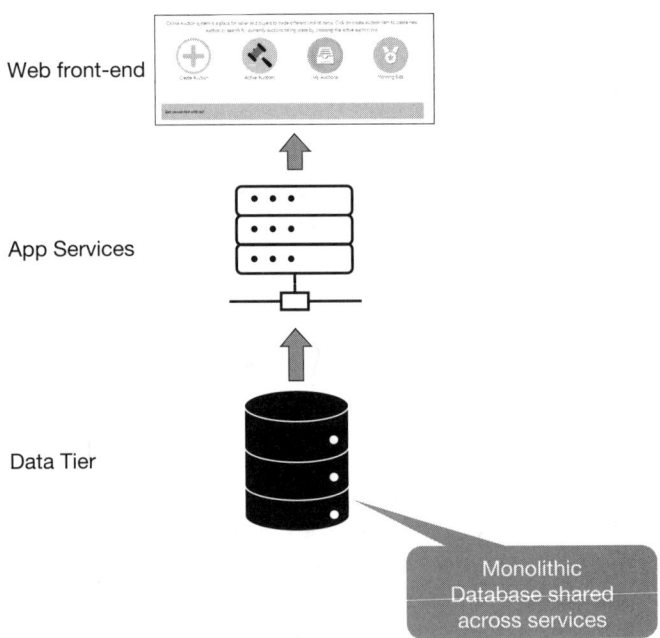

FIGURE 1-6 Monolithic database when shared across many APIs

As you can see, the data access layer in Figure 1-6 utilizes a single monolithic database that is shared across multiple services. This database also tends to be a relational database to store information, and all the app services that are operating in the middle tier perform actions directly on this single data tier.

If your application continues to grow, you might start noting issues with the scaling of your database and with maintaining this data. All the different services are recording particular information relevant to their service's operations, and it becomes very difficult to segregate this within your database.

Databases in microservices architecture

The microservices perspective is shown in Figure 1-7.

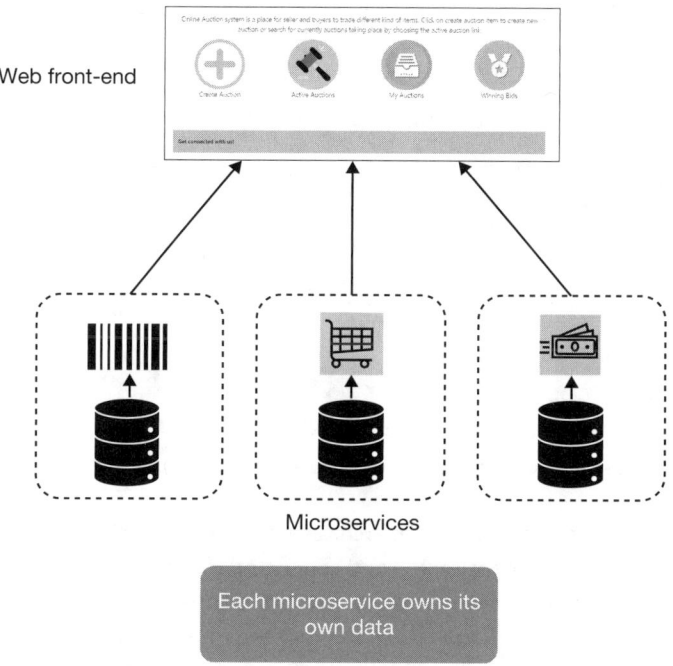

Web front-end

Microservices

Each microservice owns its own data

FIGURE 1-7 The microservices architecture

In microservices, you design with the domain or business functionality in mind. This means that you can segregate the services by business functionality, so each service can have its own particular database. Furthermore, you can specifically pick the technology behind the database of your choosing based on the need of the service. This provides even more customization, and it is a great benefit to leverage. Granted, when we abstract these services even more, we face difficulties with database consistency. Thankfully, as we mentioned earlier, there are many patterns in place to address that concern, which shows how microservices can truly make a difference in your application architecture.

Micro front-ends

When discussing the architectural features of microservices, we should also be aware of new movements and trends that continue to positively affect this paradigm, such as micro front-ends.

A micro front-end is similar to having various back-end microservices because the front-end is segmented along the same verticals. In the future, we will see more micro front-ends as we continue to evolve software architecture (see Figure 1-8).

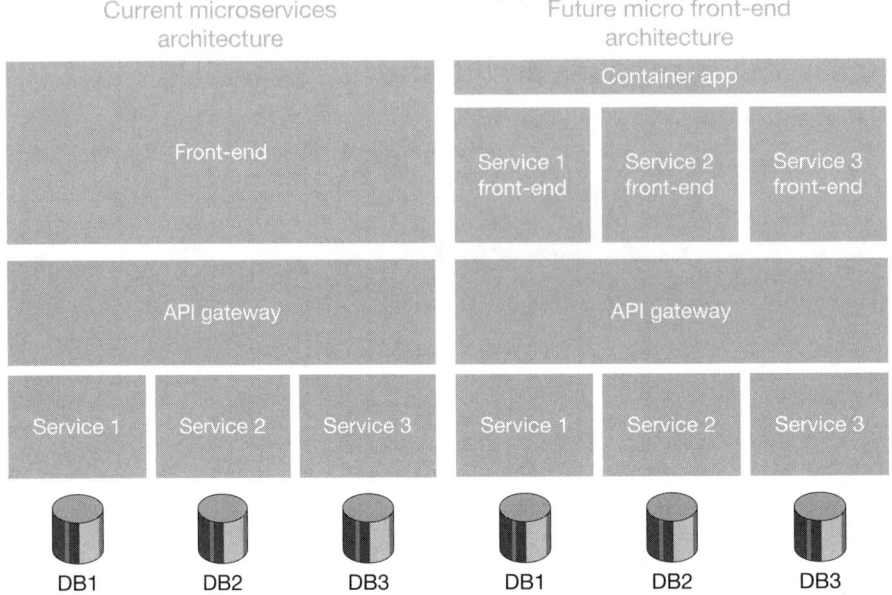

FIGURE 1-8 Microservices architecture versus micro front-end architecture

Figure 1-8 shows the current evolution of microservices architecture. When it comes to micro front-ends, we see that the front-end is further abstracted and connected to the back-end microservices over the same API gateway. This means that the UI/UX components of each of the services are segmented to those applications only. Then, all these separate components are placed together in a container app that serves as a placeholder for these HTML components with embedded data. When deployed, the components are rendered into a single front-end that is seen when the user accesses the overall application, say, through a web browser (assuming it was a web app). In traditional microservices, data components are received as JSON payloads, but in micro front-ends, data components are received as HTML components.

This componentization has similar benefits to those discussed in the next section, but this example helps illustrate its usefulness.

Let's take three teams—named Service 1, Service 2, and Service 3—operating in distinct service verticals. This means the Service 1 team can build and deploy the Service 1 full stack microservice, meaning they can build and deploy its respective data layer, business layer, and front-end. Likewise, the Service 2 and Service 3 teams can do the same, without any dependency on the other verticals.

Many of the benefits and challenges of micro front-end are strongly correlated to their respective back-end microservices. We will cover this in more detail in the next section. In this book, we use the OAS as a way to illustrate many of these concepts in practice, though we did not use this particular feature in our case study. We still want you to be aware of micro front-ends because you will likely see their underlying patterns time and time again. Different layers and components of application architecture will always be further abstracted in order to create simplified, yet richer, scenarios for developers, engineers, and more. This is something that will always grow in popularity, and is a trend to observe and track as we continue evolving our application design strategies.

Core fundamentals of microservices

A fundamental understanding of microservices is vital to understanding the examples used throughout this book, as well as the real-life case study.

First, it's important to know that microservices are autonomous, independent, and loosely coupled services that cover business scenarios and collaborate with each other to form an application. As we know, each service will have the back-end code needed to perform a particular operation. In addition, each service will most likely have an endpoint to communicate with other services for entire application functionality. Finally, these services will also have a data access layer for maintaining service-specific data in individual databases.

Benefits

We have seen some of the scenarios and understood some of the pains that enterprise architects, developers, and consultants face when it comes to application architecture. Based on our experiences, we divided these benefits into three different categories:

- Agility
- Scalability
- Technology diversity

Agility

As shown in Figure 1-9, microservices are representative of individual functions within an overall application, and as such, business functionality being encapsulated into small targeted services is one of the major points of agility. In a traditionally monolithic application, there are three layers: presentation layer, business layer, and the data access layer. In this form of architecture, all the business logic is packed together in a single layer. With microservices, this is abstracted so that each microservice is representative of one component of the overall business logic, and a single team can be responsible for building that out. In fact, this attribute alone is considered one of the major hallmarks of microservices.

This is a method to ensure logical separation of concerns. When individual teams are aligned to microservices, this tends to lend itself to more agile development methodologies and ultimately, sections of the app can be changed without regard to issues stemming from tightly coupled functions.

The second benefit in agility is that each service can evolve independently and can be deployed frequently. This is of course with the assumption that the service endpoints are well defined and agreed upon by the teams building and consuming the services. In the team dynamic setting mentioned earlier in this chapter, we noted that each service can be built out by different teams because the overall business logic of the application is further abstracted. Additionally, there isn't dependency on the deployment side either, and separate continuous integration and deployment pipelines can be set up to make changes to each team's respective service. Ultimately, one team's work on a particular service will not affect another team's work. This independence ultimately reduces the mean time to delivery from a time-to-market perspective.

This results in development teams being more agile and being able to produce at a high velocity. Ultimately, they can focus better on their individual services. Development teams are now abstracted from the internal intricacies present when working with other teams from both a technological and process standpoint, which means they can be more productive overall.

Scalability

Microservices are very scalable in relation to an application. They can scale on respective demands without affecting overall performance. This occurs because of the abstraction of business functions combined with components of the overall architecture that build out the ability to handle more traffic and requests, and it has high availability. In fact, scalability was the feature that drove the transition from SOA to microservices.

We touched on some of these components earlier. Granular services mean that requests are oriented to particular areas within the architecture. However, when we have services communicating with others for an end-to-end scenario, communication patterns and messaging channels between individual services help streamline logic to complete scenarios. For microservices architecture, we have individual communications and patterns such as publish-subscribe which makes communication easier and doesn't throttle the performance of the application.

Also, microservices tend to be containerized in enterprise scenarios and thus can be deployed efficiently and frequently with the use of infrastructure to containerize the services (like Docker), deploy with scale with Kubernetes, and be able to deploy quickly with pipelining and continuous integration and deployment.

Technology Diversity

Often, balance and flexibility are overlooked in developer environments, which gives way to some rigid architecture that has specific problems and thus requires specific answers.

Developers who work on applications that utilize microservices architectures can mix and match different programming platforms and storage technologies. This offers great flexibility for developers, architects, and consultants to design with almost limitless permutations for a final business solution. This flexibility to leverage numerous technologies is an added benefit that we see time and time again in enterprise environments.

Consequently, applications can be refactored, modernized, or experimented on with new languages and practices, one service at a time. However, the introduction of new technology into longstanding enterprise environments tends to create new problems and headaches for incumbent IT professionals. This signals the start of large-scale refactoring and modernization projects, which equate to massive expenditures in time and money. Despite this, microservices change the way we think about these initiatives and instead of generating more problems, provides solutions for IT managers.

To understand this better, suppose we started with a monolithic application a decade ago to develop a product. In the last 10 years, we find that the application code base has become huge, and the database has become monolithic, too. Today, we realized that the technology we used earlier is getting outdated and making modifications to the existing application is difficult as one change will potentially break other pieces of the monolithic application. We want to provide a modern user experience to the consumer with new technologies and applications. This is where microservices solves these problems. With microservices, because the application is segregated into set of different services, the front-end could be a micro front-end or even a basic terminal that contains some HTML pages and consumes those services to represent application features. The segregation allows teams to independently make improvements to their service with minimal impact to the rest of the application. Adding new functionality is not only easy, but the adoption of cutting-edge technologies are faster and complementary.

Challenges

We've seen some of the fantastic benefits of using microservices architecture, many of which have themes related to componentization and flexibility. On the other hand, many of these same benefits pose new challenges for your IT professionals as well.

These challenges include:

- Learning curve
- Deployment
- Interaction
- Monitoring

Learning curve

In the modern enterprise environment, we have seen the shift from a traditional team setting to a DevOps-oriented team. A traditional company might have individualized teams of developers, QA engineers, an infrastructure team, an information security team, an operations team, and database admins—all in individual silos that have to work together in a waterfall-oriented project.

Today, agile projects are the norm and teams in a DevOps setting are almost mandatory, which means that teams are composed of a smattering of people from different traditional teams to make a completely iterative team that is capable of building out and maintaining some function.

However, this means that everyone on that agile team that is assigned to building out a particular microservice must ramp up based on the different kinds of technology that are selected for the final design.

The number of technologies here poses a problem, where more options cause decision fatigue and might exacerbate the learning curve. From the developer perspective, a design choice could be made to use Java for the business logic in a particular microservice. If this isn't the primary language that developers are used to building on, then ramping up to learn the language from a syntax, semantic, and even a knowledge perspective (considering libraries available) could be a significant barrier, which can pose both time and fiscal challenges when training time and resources are taken into consideration.

This isn't a situation that is unique to developers; it can happen to any role within your agile team. Let's take an infrastructure team who is responsible for deploying the underlying components needed to build out the application. The decision on what tool to use for infrastructure-as-code (IaC)—such as Terraform, ARM Templates, and so on—and for pipelines (build and deployments, i.e. Jenkin, Azure DevOps etc.) could cause delays and expenditures out of budget for training.

Building expertise on many different kinds of technology is difficult for teams in general, and it is even more dynamic when working with the microservices architecture. That being said, good planning and an understanding of your team's strengths and weaknesses can enable you to make the right decisions to balance the time and money to make the team productive when using microservices, despite these challenges.

Deployment

Agility is one of the big benefits of microservices, particularly because builds and deployments are faster and more frequent because teams own smaller parts of an overall application. However, this bonus comes with difficulty, where one needs a strong focus and expertise in automation for faster and frequent deployments.

A manual process to address builds and deployments would be cumbersome and painstaking. Thus, automation is required to reduce this overhead. The problem is deeper here because automation of pipelines in continuous integration and continuous deployment requires specialized technologies and skills, and automated pipelines are complex to set up based on how customized your environment is. Again, this could hinder the potential pros of using microservices.

Let's understand this better by walking through a scenario. With the assumption that the code base for your service is already complete, we need to ascertain where we will be storing the code. Traditionally, tools such as GitHub provide repos and source code version control, which is useful for developers to push code, generate builds, or make improvements to the service. Primarily, we want to focus on the integration of these pushes to the repo in kicking off the workflow for a build. There are many ways to do this with your tools, but then there is a need to create the build using the artifacts created from the code and running through tests to ensure the build is working before moving it over for automatic deployment to your various environments. Each layer and step of this is complex and requires a deep understanding of multiple roles to work. This is why we see there is a high demand for DevOps and automation engineers. Furthermore, the tools that you could use for this end-to-end workflow can all vary in capability and have their own associated learning curve.

That being said, many enterprise IT professionals are observing and reacting to this need by seeking out people to help accomplish this in their environments based on incumbent talent, hired talent, and consultants they might be bringing in for these particular tasks. Microservices aside, to be successful in an agile IT world, we see this as being a scenario you cannot ignore.

Interaction

Adding on to some of the knowledge base needed to work in microservices are patterns and messaging channels. In a microservices architecture, although services are disparate and independent, the overall application functionality requires services to talk and communicate with one another for triggering of actions and functions to conceive an end-to-end cohesive workflow. Therefore, knowing messaging patterns and related technology is vital.In a traditional monolithic application, all the layers are stacked on top of each other, allowing interfacing between all parts of the application. In microservices, each function is a miniaturized monolithic application, but now it needs to communicate with several others to complete an end-to-end scenario and provide a rich user scenario/experience. Ultimately, seamless communication and independent scaling can only be achieved if services can asynchronously communicate with one another. This gave rise to technologies such as service buses and queues to help delineate communication between services.

Let's walk through an example that will reappear in detail in Chapter 6, which is how we implemented interaction with our OAS case study. The goal of this application is to provide a one-stop shop for people to auction off items, bid for items, pay once a bid is won, and subsequently receive the items they've purchased. This is a simplification of the scenario, but as you can see, with the moving parts of an application in place, several systems must talk to each other. We can see that the auction and bid services must communicate with each other, the front-end would have to communicate to all the back-end services, and the payment service must communicate with the bid services.

A service-oriented architecture has a single enterprise bus and doesn't account for all of these scenarios. Thus, a more loosely coupled architecture will require you to learn more patterns (such as request/response and pub/sub) and messaging technologies (such as Apache Kafka and RabbitMQ) to be productive. Again, this is a burden on the developer and infrastructure teams.

However, there are standard ways of learning this information, so although there is a learning curve, the information you gather will be useful in understanding how to design further architecture to handle various interactions.

Monitoring

Monitoring is a core component of many developers' day-to-day jobs. Monitoring is vital for understanding the performance of your particular application and helping identity bottle-necks, and it is used for overall troubleshooting. This is something already ingrained in the mindset of most developers. However, monitoring gets more difficult and is a strict require-ment when it comes to microservices.

> **NOTE** It is vital to log each and every detail relevant to these services, and you must be able to visualize these results. This involves the use of potential log ingestion/ collection engines, visualization tools, alerting systems, and much more. Again, this overhead needs to be considered when shifting to this paradigm.

There are several choices when it comes to monitoring, but we have found certain tech-nologies and frameworks that work particularly well for microservices. One of the most useful things to do would be to understand the native features available from the cloud provider you are using. Today, it is common for many businesses to build modern, cloud-native applications. Because of this, there are likely cloud architects on teams that are experts on the monitoring capabilities native to your cloud platform. It is extremely valuable to leverage their insights and skills to building this piece of the puzzle.

In the OAS, we leveraged Azure and so our application metrics and performance are recorded by tools such as Azure Monitor, Log Analytics, Application Insights, and much more. Whatever your platform might be, the key to addressing a potential challenge here is being informed and intelligent about the different options available to you, whether it be cloud-native, third-party tools, or open-source tools, to make good decisions for your monitoring frameworks.

Moving forward with the microservices architecture, open source, and Azure

We hope that this provides a level set for readers to understand microservices architecture as we go deeper into our case study and the decisions we've had to make with Open Source Software (OSS) tooling and using Microsoft Azure.

Moving forward, we will go into greater depth in this book, describing different concepts in detail that we've touched on in this chapter. Our case study—the OAS shown in Figure 1-9—will be the context for understanding and using these concepts as we go through this journey together.

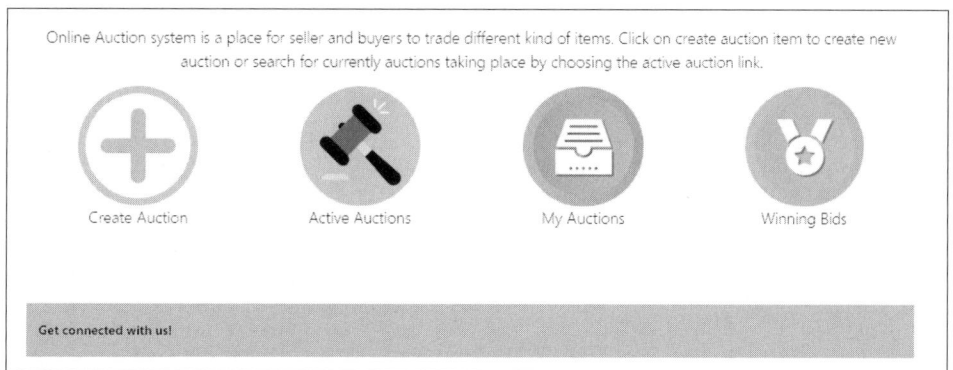

FIGURE 1-9 The Online Auction System is our real-life case study to help you understand how to use microservices architecture to build out your own modern, cloud-native applications

This book's goals

Ultimately, we hope this book will arm you—whether you are a developer, architect, consultant, or simply someone who wants to dive deeper into this topic—with the information you need to better understand the microservices architecture.

After reading this book, we hope you are sufficiently prepared to take on the challenge of building this out in a cloud-native setting such as Azure to solve your business need—whether it be through modernizing your application, migrating your application workloads, changing management, and so on. We hope that the world of microservices and open source can be the catalyst for making this a reality for you. Let's dive in!

Summary

In this chapter we covered:

- The history and evolution of software architecture from monolith all the way to microservices
- We compared different architecture models to help you understand the differences and basic inner workings of each
- We reviewed the benefits and challenges of microservices to understand the trade-offs when it comes to using this architecture

In the next chapter, we will elaborate on our real-life online auction system (OAS) case study. We will start by describing the base scenario and business requirements of the OAS and will then transition to the feature breakdown. We will discuss some decomposition principles and strategies with an emphasis on domain-driven design to demonstrate how we model our microservices.

Chapter 2

Modeling microservices— real-life case study

In this chapter, you will:

- Understand the business needs of a real-life case study, the online auction system (OAS)
- Explore the basic features of the OAS
- Learn principles and strategies for decomposing requirements into technical implementation
- Learn about domain-driven design and how to model microservices using it
- Learn some anti-patterns that should be avoided when building a microservices application

One of the key activities when designing microservices applications is to understand and model microservices based on business requirements. Various approaches, techniques, patterns, and practices are used when modeling microservices. We will discuss a few patterns and practices by using a real-life case study in which we create an online auction system (OAS) to highlight the main business processes, various techniques, and approaches used to break down business requirements and decompose them further into microservices.

In every software project, we start with the analysis phase in which we gather requirements from key stakeholders and business users. It requires a series of interviews and meetings to capture full specifications and identify business use cases.

In this chapter, we will first understand the actual business case study that we will be developing throughout this book. We will discuss the requirements and features of the application and share some concepts around decomposition techniques and practices that are important when modeling microservices.

Application requirements

The OAS is a web-based system for sellers and buyers to trade different kinds of products/items. Following are the application requirements of the OAS:

- The system should be accessible from the Internet and run on both mobile and web platforms.
- Proper user authentication and authorization is required to access application features.
- The system should support high-quality workloads.
- The application should be cloud-native and easily scalable.
- Any user who is authorized to access the system can use the following features:
  - Create an auction.
  - Each auction should be active for a particular period.
  - Auction should be auto awarded to the highest bidder once the auction becomes inactive.
  - Any user can submit bids for active auctions.
  - Users can make payments for winning auctions.

Application features

In this section, we will cover the features of the OAS.

Identity management and authentication scenarios

For identity management, the user needs to first register to access the system. The user can self-register from the website. The registration page should verify the user's email by sending the authorization code during the workflow. Once the user is registered and logged in to the system, application features can be used.

A token-based authentication is required to access the protected resources and leverage modern authentication protocols. There are various options available to implement authentication and authorization scenarios, but we used the OIDC (Open ID Connect) protocol and leveraged Azure AD B2C for identity management.

Auction management

Through auction management, registered users can sign in to the system and create auctions. Creating an auction is the first step of the business process flow. An auction contains the information shown in Table 2-1.

TABLE 2-1   Auction fields

Field	Description
Auction Name	Name of the auction.
Auction Description	Some details about the auction.
Starting Bid Value	The starting amount of the auction.
Active for Hours	Each auction will be active for a certain number of hours. The user can specify the hours while creating an auction. The system keeps the auction active until that time expires.
Auction Image	Image of the auction item.

Figure 2-1 shows the Create Auction page.

FIGURE 2-1   Create Auction

Each auction is enabled for a particular period, which is defined in the **Active For (In Hours)** field. Once time expires, the highest bidder will win the auction and make payment.

Bid management

Once the auctions are created, they become active for a particular time and users can bid for items up for auction. In order to bid for each auction, users can open the Active Auctions list and place bids for them. To win, the user needs to place a bid that is higher than the previous bid.

Figure 2-2 shows the **Active Auctions** screen.

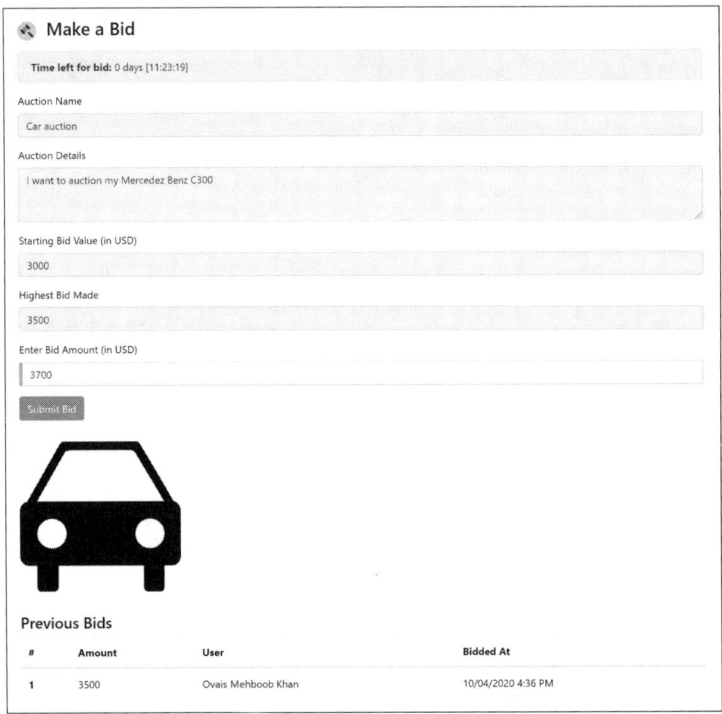

#	Name	Description	Starting Price	Last Bid Price	Remaining Time
46	Car auction	I want to auction my Mercedez Benz C300	$3,000.00		24 hrs.

Active Auctions

FIGURE 2-2   Active Auctions

The **Last Bid Price** shows the highest bid made. The user can click an active auction and place a bid on the **Make a Bid** page, as shown in Figure 2-3.

Users can specify the new Bid Amount and submit it. The **Time left for bid** shows the total time left for the bid before it becomes inactive.

FIGURE 2-3   Make a Bid

Payment management

The auction needs to be auto awarded based on the highest bid. Once the auction is awarded, the user can proceed to the **Auction Payment** screen to make payment, as shown in Figure 2-4.

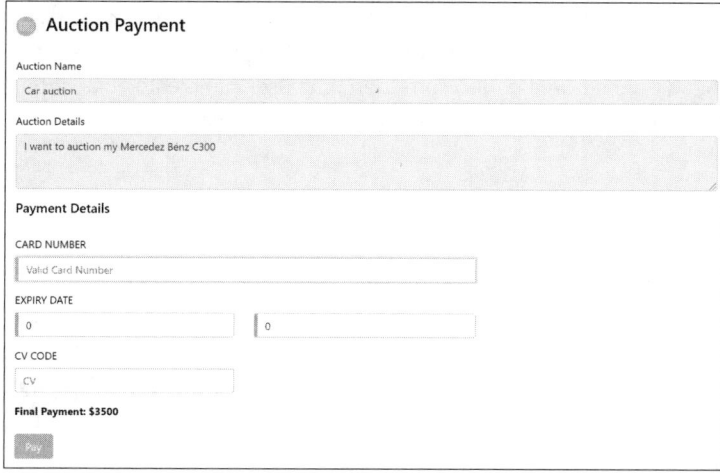

FIGURE 2-4 The auction winner uses this page to pay

Application flow

To gain better clarity on the end-to-end flow of the OAS, see the sequential diagram in Figure 2-5.

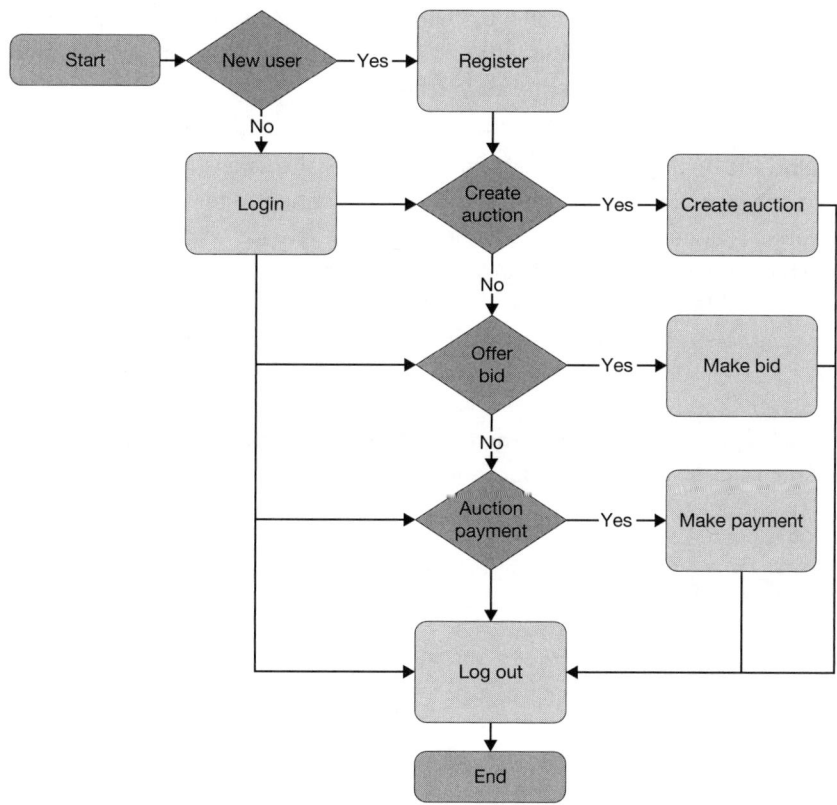

FIGURE 2-5 Application flow of the online auction system

The application flow starts when the user logs in to the system. If the user is not registered, he or she must first register. Once the user is successfully registered and logged in to the system, a list of options will be displayed on the home page. From there, the user can create an auction, view active auctions, place bids on active auctions, or pay for auctions the user has won.

Decomposition principles

When developing a microservices architecture, we need to first decompose business requirements into a set of services. When decomposing, we need to make sure that we follow these principles:

- **Service should be cohesive**   Each service should focus on a single objective and follow the *common closure principle*. The common closure principle states that the classes in a component should be closed against the same types of changes. A change in the component should not affect other components. When decomposing business requirements, we should design a service such that if any change happens to that service, it will not affect other services.

- **Service should be loosely coupled**   Loosely coupled services are independent services that have little or no impact on other services. If there are tight dependencies between services, it is more likely that a change in one service will affect other services. With loose coupling, you can independently change one service and add new functionality without considering its impact on other services and thus increase productivity and flexibility for incorporating changes and adding new functionalities.

- **Service should be autonomous**   We also have a principle where each team must be able to develop and deploy their services with minimal collaboration with other teams. If the services are capsuled and follow one business requirement, a team of developers can easily work on it independently without needing collaboration with other teams. So, when we have multiple teams work on multiple services, we should assign services based on business subdomains or business capability. So, a team on a particular domain or capability can work on one set of microservices that are part of that domain.

> **NOTE**   If we divide the services of a single domain into multiple teams, it will require more collaboration and will eventually reduce autonomy.

Decomposition strategies

There are various decomposition strategies used, though these are the most common:

- Decompose by business capability
- Decompose by subdomain

Decomposition by business capability

Business capability is one of the business features of an enterprise that expresses what an organization can do. It is a concept derived from business architecture practices. Decomposition by business capability follows *Conway's rule*, which states:

> *"Any organization that designs a system (defined broadly) will produce a design whose structure is a copy of the organization's communication structure."*
>
> MELVIN E. CONWAY

The benefit of this strategy is that it helps ensure cohesion between development teams and organization business units. See Figure 2-6.

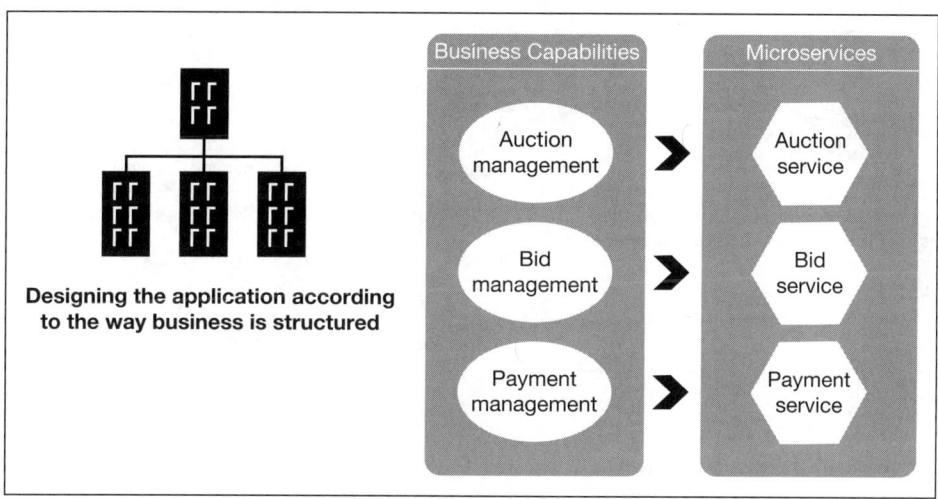

FIGURE 2-6   The decomposition of business requirements based on business capability

The business capabilities are generally stable because they define the core business of an organization.

In order to decompose the requirements based on business capability, you need to completely understand the organizational structure and the overall business of the organization. Once the capabilities are identified, we can define services based on those capabilities. Considering the OAS, we can say that the capabilities of the system are *creating auctions, submitting bids*, and *making payments*.

Decompose by subdomain

Subdomain decomposition usually requires a clear understanding of the business domains. Business domains can be evaluated on a requirement scale basis. When decomposing requirements based on subdomains, you don't need to know the overall business of an organization. Based on the requirements you receive from business users you can define domains of your system. Sometimes, if subdomains do not directly map to the current organizational structure explicitly, this can affect whether you get the acceptance of that organizational structure or change. The *Inverse Conway Maneuver Law* may be used as the basis for this change:

> "*Inverse Conway Maneuver recommends evolving your team and organizational structure to promote your desired architecture. Ideally your technology architecture will display isomorphism with your business architecture.*"

DDD (domain-driven design) is one of the most prevalent patterns used when decomposing business requirements based on subdomains (see Figure 2-7).

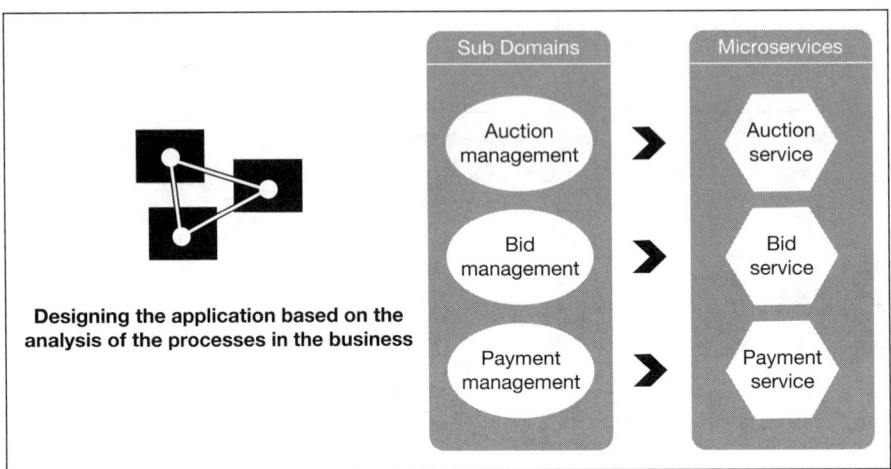

FIGURE 2-7 The decomposition of business requirements based on business subdomains

Every application has a collection of attributes, the most important of which are the amount of data it processes, the complexity of the business logic, the technical complexity, and the performance. DDD provides certain practices and principles that induce application decomposition based on subdomains.

The key difference between these two strategies is when decomposing business requirements based on business capability, the requirements are closely correlated with the organization structure, and it is easy to map and identify the processes within an enterprise to model microservices. However, if the structure of the company is not properly aligned, it can create business structural problems in the system design. On the other hand, when decomposing on the basis of a subdomain, there is often little or no organizational structure in place, and an in-depth review of the business requirements and the processes of an enterprise for the system design is needed. Nevertheless, the structure generated is more consistent with the needs of the actual business.

When decomposing business requirements based on subdomains or capabilities, it is not necessary to create a single service for each subdomain or capability. There could be more than one service to cover one business capability or subdomain. When working on a large application, we can have multiple teams who develop multiple services. If multiple teams are working on a single business capability or subdomain, it would be difficult to establish effective coordination among them, and the recommended approach for handling this obstacle is to split the team based on business capability or subdomain.

Domain-driven design

In any business application development, the domain model represents a conceptual model of the business. If the application functionality is mapped toward a business domain, it is easier to talk in its own business language known as *ubiquitous language* and tailor a model more toward that business domain.

DDD focuses mainly on the complexities of the business logic and addresses the business-related problems. DDD is not suitable for all kind of applications if the application cannot be mapped to the business domains.

DDD helps you to identity the business domains in the system and develop a system that is modeled more like a business domain. With that said, you can define models and entities in their own ubiquitous languages and better align with the core application business processes. With this approach, your application is tightly mapped with the actual business of the organization, and as your business evolves, you can update your domain models accordingly. The main concepts around which DDD revolves are ubiquitous language, bounded context, and domain categories.

Ubiquitous language

Ubiquitous language is the business language of the domain, and everyone speaks and understands the same language within that domain. This is what business users normally speak, and it also bridges the gaps between the business experts and the technical people.

Ubiquitous language allows you to keep your application code in sync with the business domain and keep your domain models and other artifacts, such as classes and tables, synced with the ubiquitous language (see Figure 2-8).

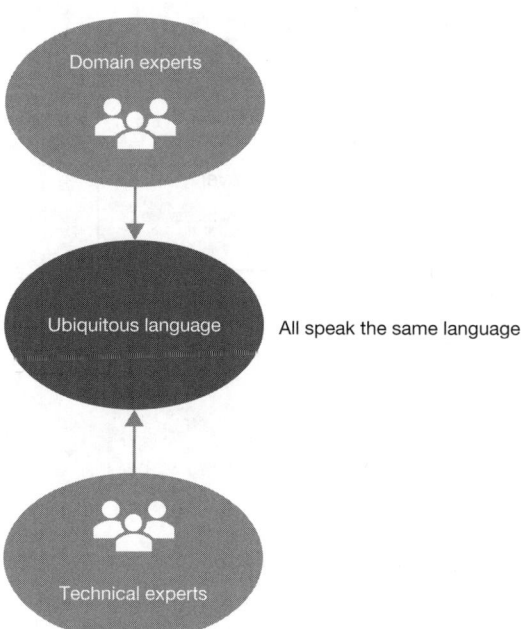

FIGURE 2-8   Technical experts and domain experts communicate in a ubiquitous language

Bounded context

Bounded context is the logical boundary of the domain. Bounded contexts provide their own ubiquitous language that is understood by domain experts. In any enterprise application, the primary domain is further classified into subdomains. For example, in the OAS, the primary domain is the Trading domain, but this can also be further classified into subdomains, such as Auction Management, Bid Management, and Payment Management.

The bounded context defines the following artifacts:

- Domain models or entities of a subdomain
- Functionality of a subdomain

Bounded context allows us to segregate the models within each subdomain as well as name the classes, domain entities, and other artifacts of the particular domain context itself.

For example, in the OAS, a user who creates an auction is a seller, whereas the user who bids on the auction is a bidder. Once the auction is awarded, the user who makes the payment is a payer.

In Figure 2-9, we can see that the Auction Management, Bid Management, and Payment Management domains are separate domains that handle the domain model of a user each in its own ubiquitous language. The bounded context contains domain entities and other code artifacts to implement specific models. With the microservices architecture, each bounded context can be represented as a single service or as multiple services.

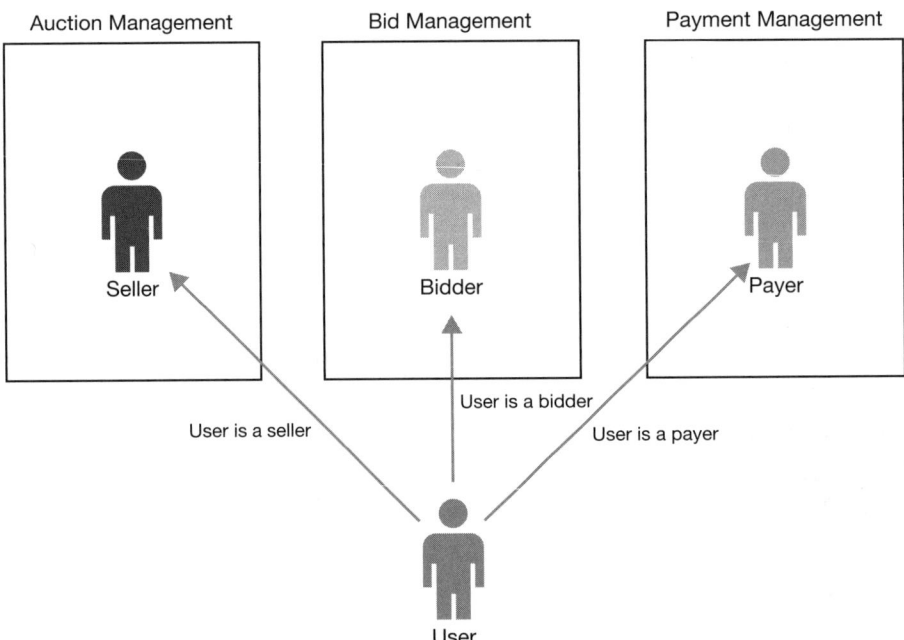

FIGURE 2-9 The user is represented as a seller, bidder, and payer in different domains

Boundaries of bounded context

Once the subdomains have been identified by the domain experts, we can define the bounded context for each subdomain in the system. In some cases, identifying the bounded context is not very straightforward, and you need to consider other factors, such as the team size or the size of an application's code. If your team is larger than 6–7 developers or if the amount of estimated application code for a particular subdomain will be large, we can create two bounded contexts for a single subdomain.

Each subdomain can have one or more bounded contexts. However, it is not likely correct to have a single bounded context for multiple subdomains. The same goes for the team of developers who are assigned to the bounded context.

> **NOTE** One team can work on multiple bounded contexts, but multiple teams cannot work on a single bounded context because this would lead to coordination issues and raise maintenance costs.

Communication between bounded contexts

In a microservices architecture, the application is divided into various segregated, independent services that are running in a separate host process. Ideally, each bounded context is represented as a single service. The communication between these bounded contexts is done through various patterns as follows:

- Direct communication over HTTP/HTTPS
- Event-based messaging using queues

We will discuss these patterns in Chapter 6, "Communication patterns."

In certain instances, a proxy service communicates to other bounded contexts in order to execute those operations. For example, with legacy applications, if you want to keep the new code separated from the existing code base, you add an anti-corruption layer that prevents entities in your model from calling entities from other bounded contexts. You add a proxy service that handles the communications between them. Instead of making the call directly, when you want to call the other bounded context, you call the proxy service that communicates to the bounded context, and the proxy service returns the response to you.

Reusing code between bounded contexts

Reusing code can be done between bounded contexts. When you are designing a microservices architecture and segregating the services based on subdomains and bounded context, there likely will be domain models that are common between different bounded contexts.

A general rule is that each bounded context should have strict boundaries, though we have seen certain patterns that negate this rule. For example, DRY (don't repeat yourself), is a software development principle that encourages the reuse of code instead of repeating and increasing redundancy. However, this seems to be an anti-pattern in the context of a bounded context. See "Anti-patterns" later in this chapter.

Each bounded context should have its own set of domain models, and its artifacts should not be reused or shared with other bounded contexts. For example, in the OAS, the seller who is part of the auction management domain can also be a bidder in the Bid Management domain. However, their entities will be separately defined under each bounded context and cannot be reused.

Domain categories

After analyzing and understanding all the business requirements and the business domains in the system, the next step is to structure them based on the following three domain categories.

- Core Domains
- Supporting Domains
- Generic Domain

Core domains

Core domains are the main domains of the business. Without these domains, we cannot run our business. For example, the shopping system, product management, order management, and payment management can be categorized under the core domain. If any of these domain services fails, it will affect the business.

In our case, we can say that the Auction Management and Bid Management are core domains of the OAS. Without these domains we cannot create auctions, and users cannot bid against active auctions. The core domains are core fundamentals of the business on which the organization's primary business depend. When decomposing business requirements, always start from the core domain because this is one of the most important and crucial domains in the system.

Supporting domains

Supporting domains are those that support the core domains. Without these domains we can run a business temporarily, but the primary business of the organization won't be affected.

For the OAS, we can set the Payment Management domain as a supporting domain because if the payment service is down, we can still use the core part of the system that is auctioning items and bidding against active auctions. However, we still need the payment service to make payments once the auctions are awarded, but for a temporary period, we can still use a major part of the system.

Generic domains

Generic domains are those that are not core to the business, but they are required to handle specific scenarios or to complete the end-to-end process flow of a particular business scenario of the system.

For example, in the OAS, each user needs to be authenticated first before using the system-protected features. So, identity management can be categorized as a generic domain. Another example is a notification service that can be used to send email notifications once the payment is made. Thus, the notification service can also be categorized under the generic domain.

Domain events

Domain events are the events that are required in your domain. Considering the OAS, we have three significant domains: Auction Management, Bid Management, and Payment Management. Each of these domains has specific domain events that are needed to perform domain-specific operations. As with the auction management domain, creating an auction can be a domain event. For Bid Management, placing a bid can be another domain event for the bid domain. Finally, for the payment management domain, making a payment can be a domain event for the payment domain. See Figure 2-10.

FIGURE 2-10 Domain events of the auction, bid, and payment domains of the online auction system

Domain events are not the system events in the application. System events are those that are not directly related to the business events. For example, system events occur when the application loads, the session starts, the session ends, when the user logs on, and so on.

The domain event is initiated when the user makes an action in the application, such as a web application, a mobile application, or any other platform. Sometimes, an event can trigger multiple events of other services, which happens when that event transaction spans over multiple services.

With the microservices architecture, this is very common because the application is divided into multiple services. However, to communicate with other services, we can either communicate synchronously or asynchronously. Availability is the drawback of making synchronous communication—which is usually done by making a direct HTTP call. If the other service is down or not available for some reason at the time the transaction is happening, it will break the data consistency across other databases because the databases are also segregated based on different domains.

Online auction system decomposition based on DDD

After going through the decomposition principles and strategies and learning much about domain-driven design, we can easily decompose the OAS based on subdomains.

When decomposing, we first identified the subdomains and drew bounded contexts against them. Then, we realized that the Auction Management and Bid Management domains are core domains because a major part of the business relies on them. Then we categorized the Payment Management domain as the supporting domain because the use of the Payment Domain is limited to only those users who have won auctions. Finally, the Identity Management domain is used to provide user registration and authentication scenarios and can be categorized as a generic domain.

Figure 2-11 shows the decomposition of the OAS based on DDD.

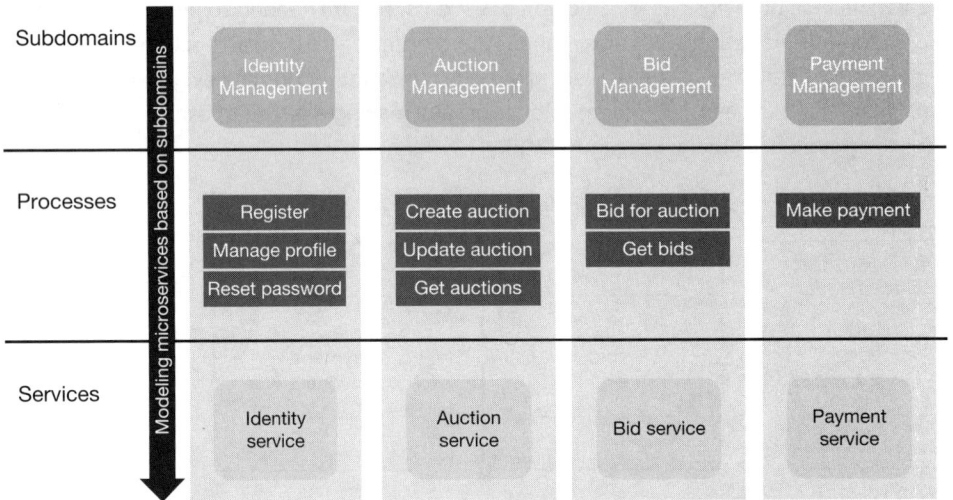

FIGURE 2-11 Decomposition of subdomains to microservices based on domain-driven design principles

Each domain has specific processes. For example, the Identity Management domain provides user registration, profile management, and password-reset scenarios. The Auction Management domain contains the Create Auction, Update Auction, and Get Auctions processes; Bid Management contains the Bid Creation For Active Auction and Obtain Bids processes; and the Payment Management domain contains the Make Payment process. Each of these subdomains can then be further mapped to different microservices.

With the microservices architecture, each service is technology- and platform-agnostic, and the database is also segregated based on business domains. Each service holds a database that contains tables or entities that are specific to its domain. However, challenges come when we need to provide consistencies across different databases.

> **NOTE** There are various patterns such as CQRS and Saga that are used to accommodate these challenges. We will discuss about these patterns in detail in Chapter 3, "Build microservices architecture."

Anti-patterns

Anti-pattern is an approach in implementing something that leads to a bad design in software development and should always be avoided. Refer to the previous section to learn about some other anti-patterns. Modeling and breaking down your business processes into fine-grained microservices is not the only rule of starting off in the right direction. There are a few design considerations that are equally important when you are building microservices.

Following are some of the anti-patterns you need to address early when you are designing and modeling microservices.

- Using monolith or shared databases with microservices
- Unnecessary fine-graining of services to deeper subdomains
- Establishing tight dependencies between code artifacts

Using monolith or a shared database with microservices

We have seen that when developing microservices architecture, many enterprises usually keep a single or a shared database to avoid complexity at the database layer. The services are fine-grained based on business capabilities or domains, but the database is a big monolithic database that is shared across all the services. This kind of implementation is an anti-pattern for the microservices architecture.

Microservices architecture is meant to have a separate database for each service in which each database should have specific artifacts (tables, stored procedures, views, and so on) based on the business domains. Changes in the underlying domain-level artifacts should not affect other services or should have a very limited or negligible effect on other services.

This kind of practice is usually seen with big enterprise solutions when they are transformed into a microservices architecture. The segregation is normally done at the services level, but because of the complexity of the database, a big monolith database is shared among all the services. This is not a good practice, so you should avoid it to the extent possible.

Unnecessary fine-graining of services to deeper subdomains

Fine-grained services always comply with the microservices architecture. If the services are fine-grained based on business domains or the capabilities of the business, they will always be aligned to the business and have a lesser effect on other services. However, when decomposing microservices, you should also consider how the transactions will be maintained.

Because the services in a microservices architecture are decomposed into several services, a single transaction could span multiple services. Achieving ACID (Atomicity, Consistency, Isolation, and Durability) transactions in microservices architecture is not quite possible, so the technique used to handle these transactions is mostly related to an event-based system, where each service communicates to the other service part of the transaction to perform the desired operation. This technique is known as BASE (basic availability, soft state, and eventual consistency). We will discuss this approach in detail in Chapter 3.

Establishing tight dependencies between code artifacts

Creating tight dependencies by sharing code is another anti-pattern. With this approach, we share the common core library between different services. In OOP (object-oriented programming), we usually have DRY (don't repeat yourself). We have seen that in a non-microservices architecture, we usually create facades or classes that are reused within other projects or components in the solution. This is a good pattern, but it becomes an anti-pattern when we work with microservices architecture.

For example, in the OAS, we have an Auction Service that is used to create auctions, list active auctions, and award auctions based on the highest bidder. With the Bid Service, we can bid on auctions. To support this pattern, we can use the existing functionality of an auction management service to read the list of active auctions in the bid service. For this, suppose if the underlying technology of both the services are same, we can add a reference to an auction service project inside a bid service and use its functionality. However, this kind of approach enables tight dependencies between the auction and bid services, so if we update anything in the auction service, we will need to recompile the bid services, too.

In microservices, all communication should be done through direct communication such as HTTP/HTTPS or through some event-driven messaging system. Keeping tight dependencies between services is an anti-pattern and should always be avoided.

Summary

Modeling is one of the core challenges when designing microservices and mapping business requirements to technical implementation. We learned the following topics:

- We began with a real-life case study of the online auction system to understand the business needs and the kind of application we will be building throughout this book.
- We studied other topics related to principles and strategies of decomposition, and we discussed domain-driven design patterns through which we eventually modeled microservices for OAS.
- Finally, we discussed some anti-patterns that are equally relevant and should be dealt with early in the microservices modeling process.

So far, we have learned about the different services we are going to develop, and we have modeled microservices based on business subdomains. In the next chapter, we will design the cloud-native solution on Azure, and we will discuss several patterns, practices, and other services and components we will be using in Azure to design the architecture.

Chapter 3

Build microservices architecture

In this chapter, you will:

- Learn about cloud-native applications and their tenets and characteristics
- Understand twelve-factor app methodology that is used with microservices applications
- Understand the overall architecture of the Online Auction System
- Get a brief introduction to the technologies used in building the application
- Learn some patterns used for distributed database architecture in microservices applications

Designing software architecture is a fundamental part of developing an application to implement the technical requirements of a project. If the architecture is robust and meets the business requirements, it makes the application compliant with the requirements in the long run and it also handles undesired scenarios effectively. When designing the architecture, we need to think out of the box and identify all the obstacles that could affect the application in terms of maintainability, availability, performance, and scalability, and we need to address them early in the architecture-designing phase.

When developing a microservices architecture, the services are independent and segregated based on the business capability. There are many factors that need to be addressed to establish an effective collaboration between services and to form an application. Furthermore, you also need to learn various patterns and practices to address some challenges. On the other hand, by making the architecture cloud-native, we can leverage many other benefits, such as scalability, availability, resiliency and minimizing the overall cost of the solution.

The OAS (Online Auction System) is a cloud-native application that is based on a microservices architecture. In this chapter, we will discuss the overall architecture of the application and showcase what Microsoft Azure provides for building the architecture and leveraging various services and components. Also, we will explore some key patterns that are essential when building a microservices application.

Cloud-native applications

Cloud-native applications are built around various cloud technologies or services that are hosted in the cloud. Microservices applications are loosely coupled and are fine-grained services that are hosted on a separate endpoint and that communicate over a lightweight protocol. Because many cloud providers offer competitive services and pricing models, making your application cloud native can improve its maintainability and scalability. When compared to cloud-ready applications, both run on the cloud, on-premises, or in a hybrid manner. For example, an application cannot be made cloud native by migrating an on-premises application to the cloud using a lift-and-shift approach over VMs (Virtual Machines) running in the cloud.

When we say that an application is cloud-native, we mean that it provides a consistent experience across all sorts of public, private, and hybrid clouds. Many organizations embrace cloud computing for flexibility and affordability. With managed services in the cloud, not only can you provision resources in minutes, but you have access to features for quickly maintaining and upgrading your services as needed. For example, when hosting microservices on an Azure Kubernetes Services, we can set certain metrics on Azure to autoscale clusters or pods based on the performance or load we experience on services. The Online Auction System is built using various open-source technologies and managed services on Microsoft Azure.

As per the CNCF (Cloud Native Computing Foundation), the official definition of cloud-native is as follows:

> Cloud native technologies empower organizations to build and run scalable applications in modern, dynamic environments such as public, private, and hybrid clouds. Containers, service meshes, microservices, immutable infrastructure, and declarative APIs exemplify this approach.
>
> These techniques enable loosely coupled systems that are resilient, manageable, and observable. Combined with robust automation, they allow engineers to make high-impact changes frequently and predictably with minimal toil.

The CNCF is an entity that supports the adoption of cloud-native applications by promoting and building sustainable ecosystems for cloud-native applications.

Main principles of cloud-native applications

Cloud-native applications are comprised of four main principles:

- **DevOps** DevOps is a collection of practices that enables people, processes, and technologies to work together and provide value to customers. It helps different roles to collaborate effectively and respond efficiently to customer needs.
- **Continuous Delivery** This is a software engineering approach to build software in a way that it can be released to production at any time. It is achieved by automating the application's build process once the change is committed by the development team,

several tests have been run, and artifacts have been released for deployment on staging or production environments.

- **Microservices**   This is a set of fine-grained services that have been modeled toward business capabilities or domains and that collaborate to form an application.

- **Containers**   Containers provide software virtualization by dividing a single server into one or more isolated containers. It provides an abstraction at the application layer that packages application code and dependencies together. Containers take less space than a VM (virtual machine) and share resources, such as CPU and memory.

Figure 3-1 shows the key components of a cloud-native application.

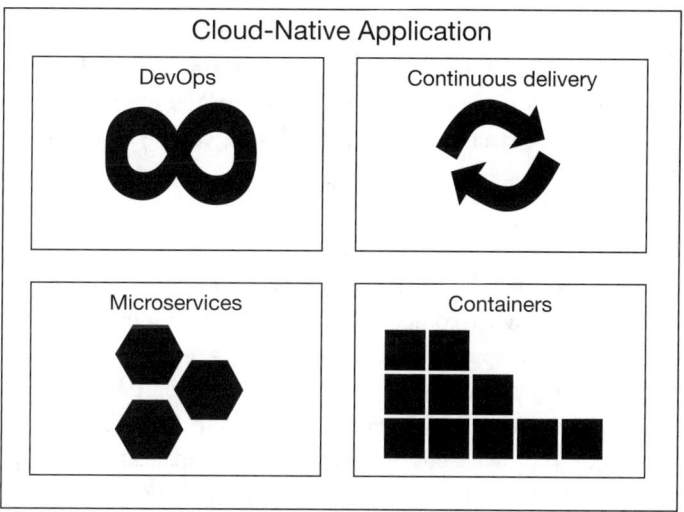

FIGURE 3-1   Key artifacts of a cloud-native application

Characteristics of cloud-native applications

Cloud-native applications offer many benefits for microservices applications:

- **All cloud-native components**   Cloud-native applications embrace cloud technologies to build and host applications in the cloud. These applications drive extensive use of PaaS (Platform as a Service) services in the cloud. Services can be provisioned in minutes and offer a competitive SLA (Service Level Agreement). Cloud-native components also provides many monitoring options to help administrators see how the services are performing and either scale out or scale up, depending on the requirement. It's best to make the microservices architecture cloud native, so you can leverage faster spin-up of resources and have the ability to scale out or scale up on special circumstances; you can even scale down to save costs.

- **Faster release to market**   Cloud-native applications offer faster application delivery to customers by using DevOps to automate software releases via CI (continuous integration) and CD (continuous deployment). The automation can be achieved at the deployment level or by upgrading the application version, but it can also be used to set up the whole infrastructure using an IaC (Infrastructure as Code) approach.

Provisioning and maintenance with ease

The cloud platform offers services at the IaaS (Infrastructure as a Service) level, PaaS (Platform as a Service) level, and SaaS (Software as a Service) level (see Figure 3-2). IaaS has some maintenance in terms of applying upgrades or software patches, but no hardware or infrastructure cost is needed to purchase the hardware and set up the infrastructure. For example, a VM running on Azure is IaaS.

With PaaS, everything is managed and maintained by the cloud provider where you, as a developer, only need to focus on the application code. Some examples of PaaS are Azure App Services, Azure Kubernetes Services, and Azure SQL. With SaaS, you don't even need to build the service or application; instead, you can subscribe for the service and usage. Office 365 is a good example of SaaS.

Cloud-native applications offer high maintainability by providing competitive features in the cloud to maintain applications with ease. Whether it's a container-based application deployment or spinning up VMs, all can be done with ease. With Microsoft Azure, we have the option to maintain cloud resources, either from a portal or Azure CLI or by setting up certain metrics to automate the scaling.

Cost-effective solutions

Cloud-native applications utilize cloud services that usually provide a competitive pricing model as compared to running everything on-premises. When compared to a standard on-premises deployment, maintaining the infrastructure, scaling out applications, or scaling up the resources can be expensive in terms of time, effort, and cost. However, by estimating the cost of an on-premises solution properly, you can realize the savings in the cloud. Over a cloud platform, the resources can spin up in seconds and can be configured to autoscale based on certain metrics that can be configured in minutes. With PaaS services, there is not much need for maintaining infrastructure, too.

Availability and scalability

Generally, all cloud providers provide numerous managed services where each resource has a specific SLA (Service Level Agreement) that you need to know before provisioning it. With cloud-native applications, because the applications are hosted in the cloud, we usually get higher availability and scalability. Setting up something similar to what the cloud offers on-premises is not easily possible and requires a huge investment.

Resiliency and reliability

Resiliency and application reliability are among the top characteristics of cloud-native applications. With a microservices architecture that follows the modern cloud-native approach, resiliency and fault tolerance are built into applications. See Chapter 6, "Communication patterns," for more details.

> **NOTE** Because the services communicate over a lightweight protocol such as HTTP, resiliency must be embedded in the application to enable reliability. Thus, retry and circuit breaker patterns are two of the primary patterns used to enable resiliency.

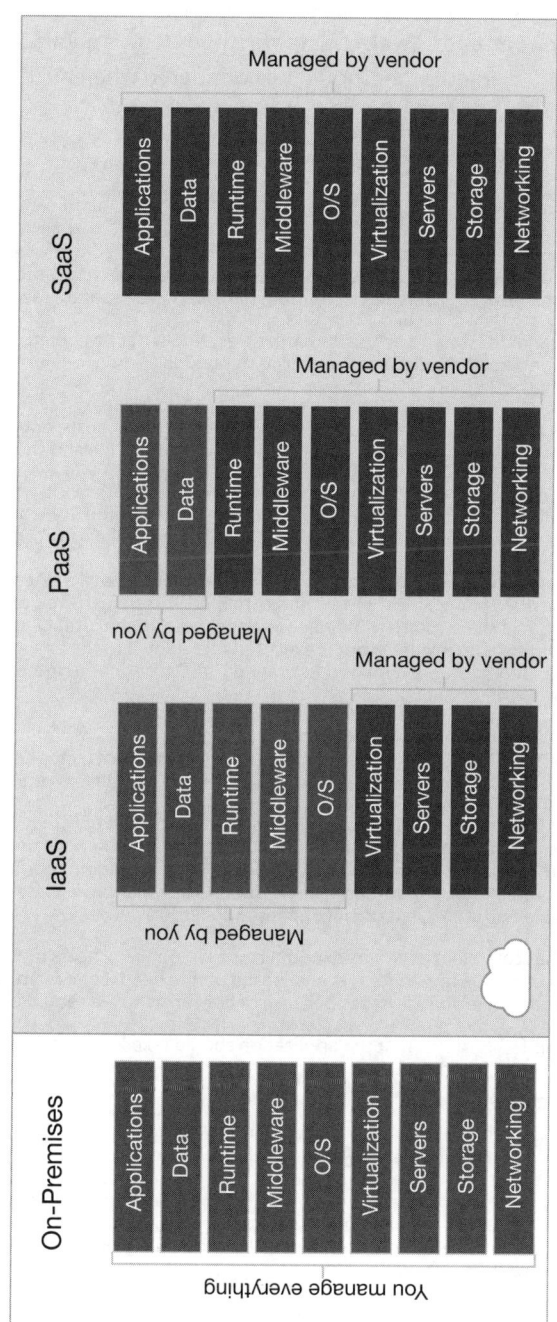

FIGURE 3-2 This is a comparison between on-premises, IaaS, PaaS, and SaaS. The concept for this figure was taken from *https://docs.microsoft.com*

Twelve-factor app methodology

In the current era, most applications are based on web and server-side technologies. The key application components revolve around web services for backends, and prevailing client-side frameworks are used for front-ends. Twelve-factor methodology is applied to web-based applications. It provides 12 elements that should be considered when building applications that use this methodology. Table 3-1 covers the 12 factors:

TABLE 3-1 Twelve-factor app methodology

Factor	Description
Codebase	This factor states that there should be some version control uses for the application codebase. For the OAS, we have used Azure DevOps with Git to store the application codebase. Each service or component resides in a separate repository.
Dependencies	This factor states that the dependencies should not be copied directly to the project codebase. It should be added using dependency-management tools such as NuGet, NPM, and so on.
Config	All the key data, such as connection strings, services ports, endpoints, and others, should be defined in the application's confirmation file and should not be hardcoded.
Back-end services	The back-end services should be loosely coupled with front-end services and use various communication mechanisms (synchronous and asynchronous) to communicate with applications using HTTP and message broker protocols and techniques.
Build, release, and run	Applications should have a separation between the build, release, and run stages. In the build stage, the application is compiled into a binary and drops the binary artifact in a shared folder. The release stage picks up the artifact and assigns the release number. The release script is executed to deploy it to the environment. For the OAS, we used Azure DevOps and leveraged CI/CD (continuous integration/continuous deployment).
Processes	This states that the application should be executed as one or more stateless processes. The microservices architecture adheres to this factor because the applications are usually stateless services running as different stateless processes.
Port binding	This factor states that the application or service port on which it is running should be configurable and stored in a configuration file. Each service runs as a separate host process and is bound to a specific port. The port binding should be configurable and stored in the configuration file. In the OAS application, each service port is configurable and stores in its own configuration file.
Concurrency	The scaling out of services should be based on the process model, which means services can be horizontally scaled based on the need. For example, if a bid service has many performance issues because of concurrency, we should be able to scale out.
Disposability	This factor states that the application should take less time to start and should offer a graceful shutdown. With a microservices application, there are fine-grained services that offer fast startup times.
Dev/Prod parity	This factor states that your development environment should be similar to the staging or production environment so there will be fewer deployment issues. Containerization is one of the paradigms that support this factor because the application runs inside the isolated environment within the container itself, and there are usually no dependencies required at the environment level. Apps running inside the container on a development environment will work similarly on the staging and production environments, too.
Logs	This factor states that logs should be treated as event streams where you can keep your logs. Various logging frameworks are available, but for the OAS, we ended up using Application Insights to send custom logging telemetry.

Factor	Description
Admin process	The admin process states that all the admin-related tasks should be automated and performed using scripts. With the OAS, we have enabled IaC (Infrastructure as Code) using terraforms and have written some scripts to be used with Azure DevOps CD pipelines to automate the deployment.

The online auctioning system (OAS) architecture

The OAS is a cloud-native application that is built on Microsoft Azure. Microsoft Azure provides several managed services on Azure that can be provisioned within minutes and provides high availability and scalability options. Figure 3-3 shows the architecture diagram of the OAS that are used by the managed services in Microsoft Azure to design the overall solution.

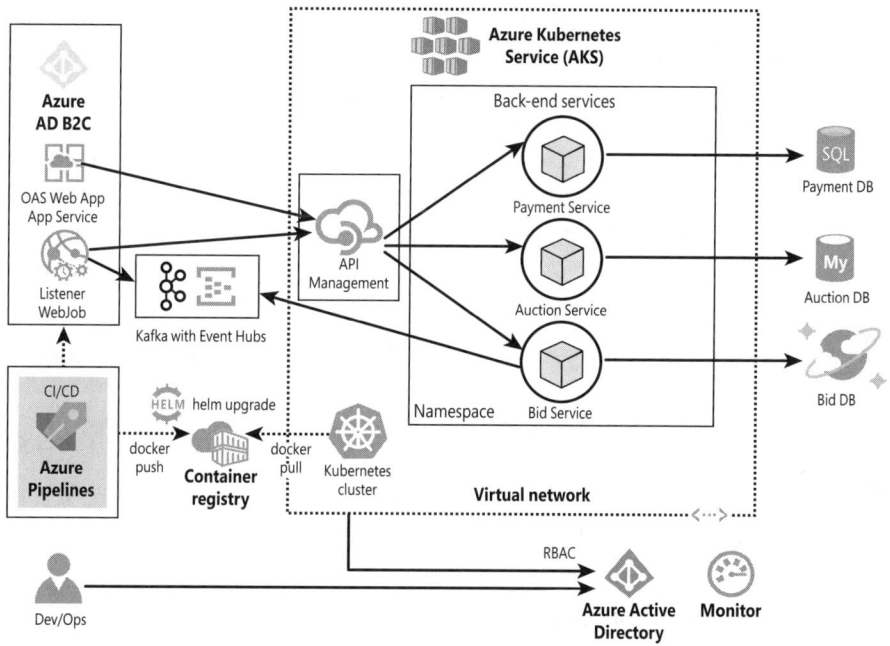

FIGURE 3-3   High-level architecture of the Online Auction System

In Figure 3-3, starting from the top left, three services are shown: Azure AD B2C, OAS App Service, and Listener WebJob. Azure AD B2C is used to provide identity and access management to the website. Azure AD B2C can be provisioned from Azure and provides a B2C solution for allowing users to self-register to the Azure AD B2C tenant. In this case, the admin does not need to create user accounts; instead, the user can visit the site and register. We integrated

Azure AD B2C with the front-end application to provide user authentication and authorization scenarios using modern authentication protocols such as Open ID Connect (OIDC). The front-end application is hosted as an Azure App Service. Azure App Service provides a PaaS model in Azure to deploy the application without considering the server-level configuration that is usually required when deploying it on-premises or on a VM in the cloud. With App Service, you can concentrate only on your application code and deploy it using a command-line option, Azure portal, or many other options.

The Listener service runs as a background service using WebJobs in the same App Service plan and is used to listen to events to provide integration with other services by making calls through APIM (API Management). Azure WebJobs enables you to run a program or script in the same instance as App Service. To enable event-driven messaging, we have used Azure Event Hubs with the Kafka protocol. Refer to Chapter 6 to learn more about Kafka.

In the center of the diagram shown in Figure 3-3, AKS (Azure Kubernetes Services) provides an easy way of deploying and managing containerized applications in the cloud. It can be provisioned within minutes and offers a serverless Kubernetes experience. All the microservices—the auction, bid, and payment services—are running as Docker images and are hosted inside the AKS cluster. Web front-end applications cannot directly access these services because there is an APIM in front of each service. APIM acts as a reverse proxy for all the services and provides a single front-door proxy endpoint to access these services. The databases are also managed databases in the cloud and are used by the services. To store Docker images, we are using ACR (Azure Container Registry), which enables you to create, store, and manage container images and objects for all kinds of container deployments in a personal registry.

> **NOTE   Azure DevOps CI/CD pipelines**
> To build and release services to an AKS cluster, we utilized Azure DevOps CI/CD pipelines. Azure DevOps provides various plugins for open-source technologies, including the ability to build a Java application or Node.JS and deploying it to Kubernetes, as well as using Helm to set up the whole Infrastructure-as-Code.

Lastly, for monitoring, we have used Application Insights and integrated the App Insights SDK with all the services to send out custom telemetry to the Application Insights resource. We also used Azure Monitors to monitor the performance of our containers running in AKS.

Representation of Azure Kubernetes cluster nodes, pods, and services

We set up a three-node Kubernetes cluster on AKS, where each node has three pods, and each pod is running a container service inside it (see Figure 3-4).

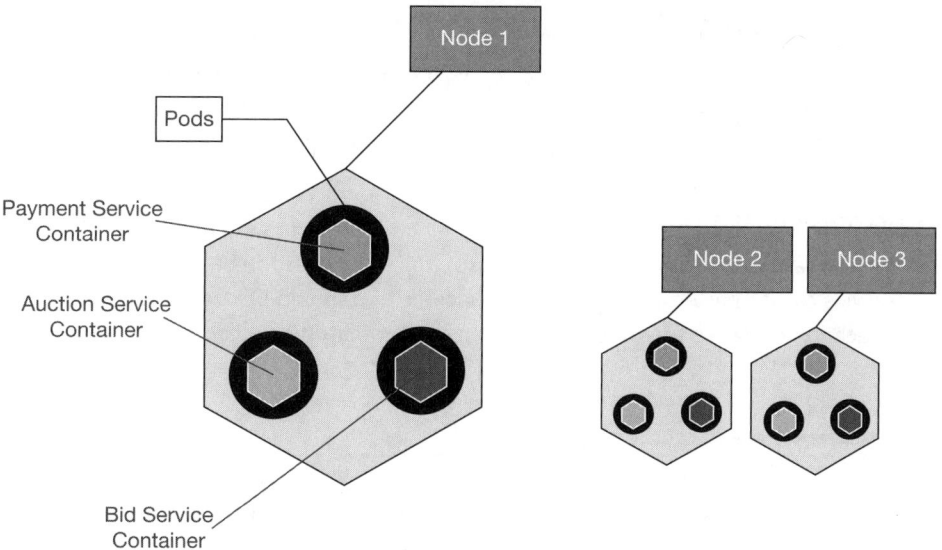

FIGURE 3-4 A representation of the nodes, pods, and services hosted in AKS

Technologies used

In the architecture diagram shown in Figure 3-4, we saw many of the open-source technologies used to develop the architecture. This section briefly introduces these technologies.

Front-end technology

SPAs (single-page applications) are a widely adopted, web-based model for developing rich applications that run on the web. There are lot of client-side frameworks available, though we used Angular for developing the front-end part of the OAS.

> **NOTE** **Angular**
>
> There are many client-side frameworks available that can be used to build SPA applications, such as Angular, ReactJS, and Vue.JS. However, we ended up using Angular for the OAS because it is backed by Google and has better community support. Angular provides a complete development framework for both web and mobile applications. Refer to *https://angular.io/* to learn more about Angular.

Technologies used for building microservices

Choosing the right technology when building microservices is important. Many compelling services and framework models offer a competitive edge for building microservices. One way to choose the best technology for your scenario is to follow the Microservices chassis pattern.

Microservices chassis pattern

A chassis pattern can be thought of like a car chassis, which provides the base frame for your car. Microservice chassis patterns provide some concerns that should be considered when choosing a technology. Make sure it supports all these cross-cutting concerns, as shown in Table 3-2.

TABLE 3-2 Microservice chassis concerns

Concern	Description
Externalizing configuration	Outsourcing the application connection strings, secret keys, endpoints, and so on, to a configuration file or some other external vault. If you are modifying any of the key values, you should not need to recompile the service.
Logging	Supports the logging framework or integration with any logging framework to get diagnostics information about the service or to troubleshoot any reactive issues.
Health checks	You should be able to expose certain endpoints to see the health of the service.
Exception handling	Provide proper exception handling mechanisms to catch errors and log.
Security	Should provide an option to secure the service or integrate with other authentication/authorization providers.
Metrics	Measurements that provide information about how the application is performing.
Distributed tracing	Instrument services that assign a unique identifier to each external request.
Connect to third party services or databases	The ability to connect with third-party services or perform database operations.

There are many frameworks and languages available today to develop microservices, including these:

- .NET Core
- JavaSpring boot framework and JavaSpring Cloud
- Node.JS—Express JS, Molecular, Seneca
- Go kit
- Micro
- Gizmo

We used .NET Core, JavaSpring boot framework, JavaSpring Cloud, and Node.JS Express for building the microservices used in the online auctioning service. Figure 3-5 shows the representation of the technologies used when building microservices for auction management, bid management, and payment management services.

Auction Service
Node.JS 10.13.0
Framework: Express JS

Bid Service
Java 1.8
Framework: Spring Boot

Payment Service
.NET Core 3.1
Framework: ASP.NET MVC

FIGURE 3-5 Languages, versions, and frameworks used to build different services

Choosing different technologies for each service does not mean that it cannot be used with other services. The primary reason is to showcase how different technologies can be used and what challenges one can face when building microservices-based applications. Technology agnosticism is one of the primary advantages of working with microservices.

Cloud technologies

This section specifies the Azure-managed services used in the OAS.

Azure App Service

There are many options for deploying web applications on Azure. You can deploy on a VM, use Azure App Service, or containerize and run them on AKS. We chose Azure App Service to host the front-end application based on Angular because Azure App Service is a PaaS (Platform as a Service) model and with PaaS, there is no maintenance needed for the underlying VM where the application is hosted, and you don't need to set up the web server. Instead, the focus is more limited to the application code itself where developers can only concentrate on the application code, rather than thinking about the deployment configuration or installing run-times on the hosted server. Another benefit is that the pricing model is more competitive than the IaaS model, where you need to provision and maintain VMs and configure web servers and other hosting-related dependencies.

With Azure App Service, we can quickly build, deploy, and scale web applications on Azure. It supports all the common application-development frameworks, such as .NET, Java, Node.js, Ruby, PHP, and Python. Azure App Service also supports hosting and deployment of Docker containers. So, if you are running a local Docker image that you want to run on Azure, using Azure App Service is one way to do it.

Azure WebJobs

Azure App Service also provides a feature that runs background tasks under the same App Service Plan. Because our Listener WebJob is an ASP.Net core-hosted service that runs in the background, is integrated with Kafka, and provides integration by making calls to other services, we chose Azure WebJobs so we could host it on Azure.

WebJobs supports the following types of files:

- Windows commands (.cmd, .bat, .exe)
- PowerShell script (.ps1)
- Bash script (.sh)
- Java (.jar)
- JavaScript (.js)
- Python (.py)
- PHP (.php)

There are two kinds of Azure WebJobs: continuous and triggered.

- **Continuous WebJobs**   These kinds of WebJobs start immediately as they are created. If your requirement is to host a service that needs to run continuously, you have to build logic inside your application itself to run it for an indefinite period. This can be accomplished by using some looping techniques, threading, and so on. Continuous WebJobs can also run on all the instances of the web app and scale out to the number of instances on which your web app is running.

- **Triggered WebJobs**   These webjobs need to be triggered manually or by scheduling them through a CRON schedule. When configuring the triggered WebJob, you can specify the CRON expression that holds a `Timespan` value to run the application at that specified time. The CRON Expression has the following format:

```
{second}{minute}{hour}{day}{month}{day of the week}
```

Table 3-3 provides some examples of CRON expressions.

TABLE 3-3   CRON expressions

To schedule	Expression
Runs daily	0 0 0 * * *
Runs every 30 minutes	0 */30 * * * *
Runs every hour	0 0 * * * *
Runs every 3 hours	0 0 */3 * * *
Runs every Sunday	0 0 0 * * 0

In the OAS, the Listener job contains built-in logic to run the application on an indefinite loop. Because we don't need to configure it as a triggered WebJob, we are running it as a continuous WebJob. A job starts automatically once a new message arrives at the Kafka topic.

Azure Event Hubs

To provide asynchronous communication between services, we used Azure Event Hubs with the Kafka protocol. In our case, when the bid is made, the auction service publishes a message to the Kafka topic for a bid named *bidtopic*, and then the listener service that continuously listens for messages from bidtopic makes an HTTP PUT request to the auction service to update the database with the latest bid information.

Azure Event Hubs is very easy to set up and can be provisioned in seconds. It is a fully managed PaaS service with minimal configuration overhead. It supports the Kafka protocol, so if you have applications that are using libraries to publish or subscribe messages to/from Kafka, you can seamlessly use Azure Event Hubs. Table 3-4 shows some of the key components of Azure Event Hubs.

TABLE 3-4 Azure Event Hubs components

Component	Expression
Event producers	Objects that send data to Event Hubs. These can be applications, APIs, or any other service. The events can be published using HTTP, AMQP, or Kafka protocols.
Partitions	You can partition the stream where each consumer can read a specific partition of the message stream.
Consumer groups	Consumer groups represent a view of the Event Hub. Each consumer application has its own view of the event stream. The view is represented in the state, position, and offset that helps each consumer read the message at his or her own rate and offsets.
Throughput units	Throughput units show the capacity of the Event Hubs.
Event receivers	Objects that read data from Event Hubs.

Azure API Management

With a microservices-based architecture, the system is divided into various APIs. Exposing all the APIs directly to client applications or consumers is not a preferred option. Each API is hosted as a separate service and is listening on a different endpoint having a different IP address and port.

The APIM (API Management) offers a hybrid, multi-cloud management platform to control your APIs. Any API on any platform can be added, and APIM supports various specifications, such as Open API specification, WADL, and WSDL. Once the APIs are added into the API Management platform, you can set up policies to control the APIs at the inbound, back-end, and outbound levels.

For example, if you want to validate the incoming token for every request that is made to the API, you can use and configure the `validate-jwt` policy template in inbound policies and validate all the tokens being passed throughout the requests being made to the back-end API. Otherwise, if you keep this logic at the back-end API level, you have to implement it on every API, which requires a lot of effort. These features allow us to develop enterprise-grade APIs that can be consumed at a global scale.

Azure AD B2C

Azure AD B2C provides an identity-management solution for applications that allows users to self-register to the system. The OAS is a public-facing system, where any user can register. Once the user is registered, he or she can use features such as creating new auctions, bidding against active auctions, and making payments once auctions have been awarded. With Azure AD B2C, we can set up an application to outsource authentication and authorization scenarios and leverage out-of-the-box user interfaces for user registration, profile editing, and much more.

An Azure AD B2C tenant can be provisioned from the Azure portal and you can create various user flows to register users, edit profiles, reset passwords, and more. Once the user is registered, the system redirects the request to the Azure AD B2C tenant user flow and shows the web page with the fields required to register. While setting up the user flow for registration or sign in, you can configure the fields part of the registration process and the fields part of the token itself as it relates to claims. Integration with your web front-end application is easy by leveraging the MSAL (Microsoft Authentication Library), which is available for many languages and frameworks.

Azure Kubernetes Services

AKS (Azure Kubernetes Services) provides a fully managed, secure, and highly available container orchestration solution on Azure. AKS is easy to set up in the cloud, and it takes just minutes to provision or scale resources. You can easily deploy and manage containers and accelerate deployments by using CI/CD pipelines of Azure DevOps. Figure 3-6 illustrates an AKS cluster.

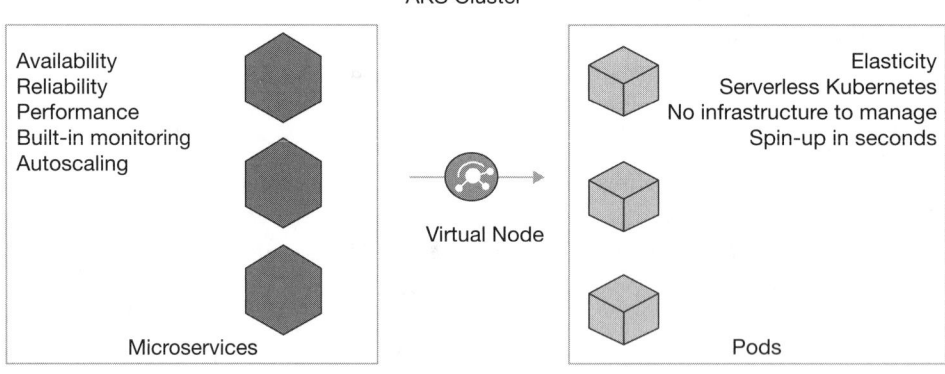

FIGURE 3-6   AKS cluster

AKS offers elastic provisioning, which enables you to configure autoscaling for your cluster or pods based on your needs. The compute capacity can elastically increase in seconds without worrying about the underlying infrastructure. We have used AKS to orchestrate the containers for all the microservices developed for the OAS.

Azure Container Registry

To deploy images to AKS, we need to first store them somewhere in the registry. ACR (Azure Container Registry) offers a private registry to store docker images. ACR is a managed resource that can be provisioned in Azure within seconds. It can be used with multiple environments such as AKS, Azure Red Hat OpenShift, and even services like Azure App Service, Machine Learning, and Batch.

You can tag and push local Docker images to ACR. By using Helm or Kubectl commands, the images from ACR can be deployed to AKS. To learn more about the cycle of pushing images to ACR and then deploying them to AKS, refer Chapter 5, "Microservices on containers."

Azure DevOps

ADO (Azure DevOps) plays an important role in any software development cycle such as team planning, following agile methodologies, versioning source code, setting up build and release pipelines for automating builds and deployments, creating test plans, and running tests on the cloud.

With a microservices-based application, the application is divided into several services, and deployment is one of the important factors. With ADO, we can set up CI/CD pipelines that help with building and deploying the container images on AKS with a one-time configuration. Imagine, if you want to update services to the AKS cluster manually, you have to run several commands to first tag it to the ACR (Azure Container Registry) and then deploy it to AKS. With ADO, the CI and CD pipelines need a one-time configuration, and the rest of the deployments can either be automated or done with a single click. We have used CI/CD pipelines to deploy images to AKS, but they also can be configured as IaC (Infrastructure as Code) to set up the whole infrastructure on Azure using a Terraform template.

Azure Application Insights

Azure Application Insights is a cloud resource that can be integrated with any application running on-premises or in the cloud. Azure Application Insights is used to monitor your applications. It can detect how the application is performing and provide you with real time insights. It also offers a powerful analytics tool to diagnose issues and understand how the application has been used.

Azure Application Insights can be integrated with various languages and provides native SDKs for Node.JS, Java, .NET, and Python. You can also use Application Insights Status Monitor V2 to configure Application Insights to your .NET applications running on a server without even modifying the application code. However, in cases where you need to send out custom telemetry, you need an SDK.

Azure Application Insights is not limited to web applications. Desktop applications, mobile applications, and even background services can use Azure Application Insights for monitoring.

In the OAS, we have integrated Application Insights with all the microservices that are built on Java, .NET Core, and Node.JS. The front-end application is built on Angular. To see how this has been integrated, see Chapter 10, "Monitoring microservices."

Azure Monitor

For AKS and Cosmos DB monitoring, we used Azure Monitor which provides a central place for monitoring all your cloud and on-premises resources. With an advanced analytical engine, you can reveal hidden patterns in your data and query results based on different constructs.

Azure Monitor collects the telemetry from various sources and aggregates the telemetry in a log data store that is optimized for cost and performance. With Azure Monitor, alerts can be easily set up on various metrics to take actions when that metric condition is met. For example, you can scale out the AKS node when the CPU threshold reaches 70 percent.

The telemetry collected from various sources can also be visualized into workbooks, views, dashboards, and Power BI.

Distributed database architecture

Microservices are segregated, and generally, each service holds the domain-specific informa-tion in its own data store unless there are any exceptions. Today, databases are not limited to a relational database model, and NoSQL and non-relational databases have been rapidly adopted. With a relational database, tables are structured such that a single domain object sometimes spans multiple tables. You could consider a relational database scheme where sev-eral bids can be made against a single auction object and a one-to-many relationship between auction and bid tables is enforced. On the other hand, NoSQL databases allow you to keep the whole transaction as one document in JSON or XML formats. Also, you can use REST OData style queries to filter out results or search documents. Choosing the right database technology is challenging because of the nature of the data being stored for each microservice.

Transactional data model

The data can be termed as "transactional" when it holds the transactions made in the sys-tem. We normally use a relational database for transactional data because it supports ACID (atomicity, consistency, integrity, durability) transactions. Another benefit is that with relational databases, the data is normalized and allows quick updates or retrieval. In the OAS, we used relational databases for both auction and payment management transactions.

Transient data model

Transient data is short-lived data that contains user activity, logs, user sessions, and so on. This type of data can be stored in a NoSQL database because it consists mostly of insertions to record the events taking place in the system and there are fewer updates or modifications to the data. NoSQL databases are usually fast when it comes to insertions and record retrieval because it supports a schemaless architecture. There is no validation being done when the data is inserted.

In the OAS, we chose a NoSQL database for the bid service to store bid-related information, and we used Mongo DB API for Cosmos DB in Azure.

Polyglot persistent architecture

The architecture can be termed as polyglot persistence if it is based on multiple database technologies. The idea behind using multiple database technologies is to embrace performance, scalability, and cost. However, it increases challenges with data consistency, fragmentation, and management of your data.

Figure 3-7 depicts different database technology used for each microservice. For example, the auction service uses a MySQL database to store auction-related information. The bid service uses MongoDB API for Cosmos DB to store bid information and uses SQL Server for payment information.

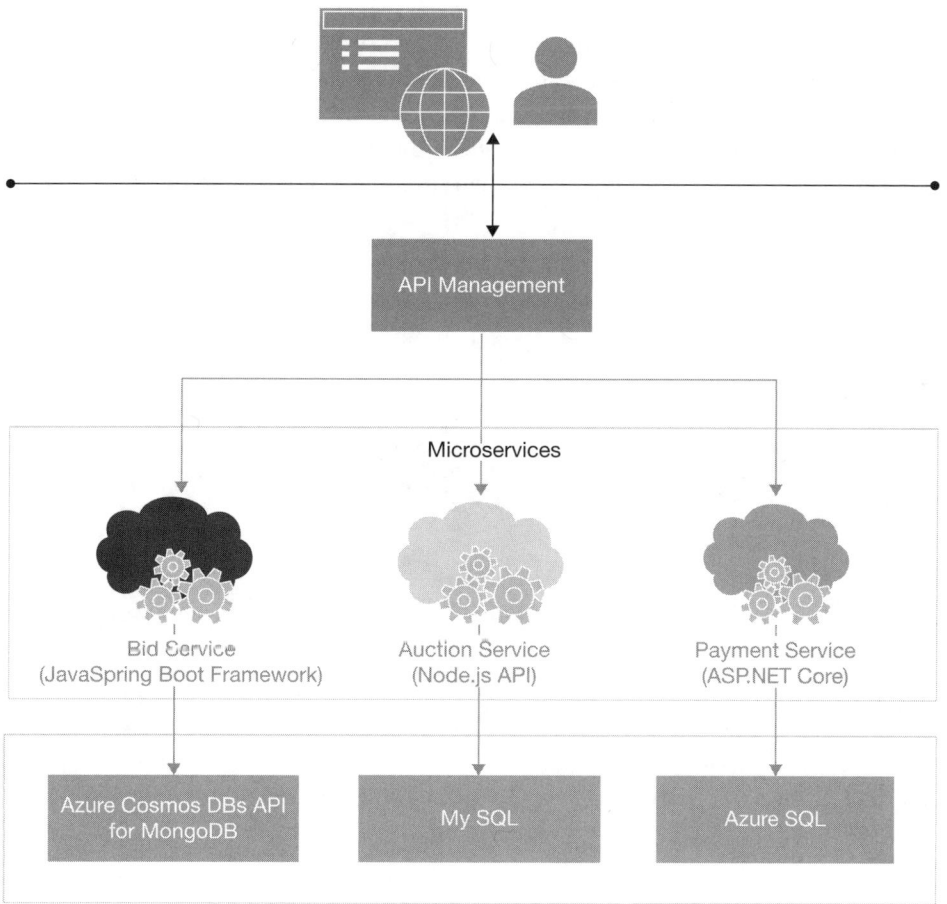

FIGURE 3-7 A polyglot persistent architecture using different database technologies for each service

In the OAS, all the databases are managed databases in Azure, and the services running inside the containers connect to these databases using its respective connection string.

Patterns in distributed databases

There are challenges when reading or accessing the data stored in multiple databases based on various polyglot technologies. There are scenarios where a transaction spans multiple services and maintaining consistencies across all the databases for part of a single transaction is challenging. In this section, we will explore some patterns that can be used when working with a distributed databases architecture.

Direct HTTP call

Direct HTTP call is considered to be an anti-pattern and should always be avoided because it creates a direct dependency on other services. Figure 3-8 illustrates an eCommerce system that contains three services: shopping service, product service, and pricing service. Each service holds separate databases to store domain-specific information. When a user buys items from the eCommerce site, the site makes an HTTP POST request to the shopping service. Prior to saving the request to the shopping service database, the service needs to validate the product quantity and read price information. Using a direct HTTP call pattern, the shopping service can make a direct request to the product and pricing services to read the product and pricing information and then save the information to the shopping service database.

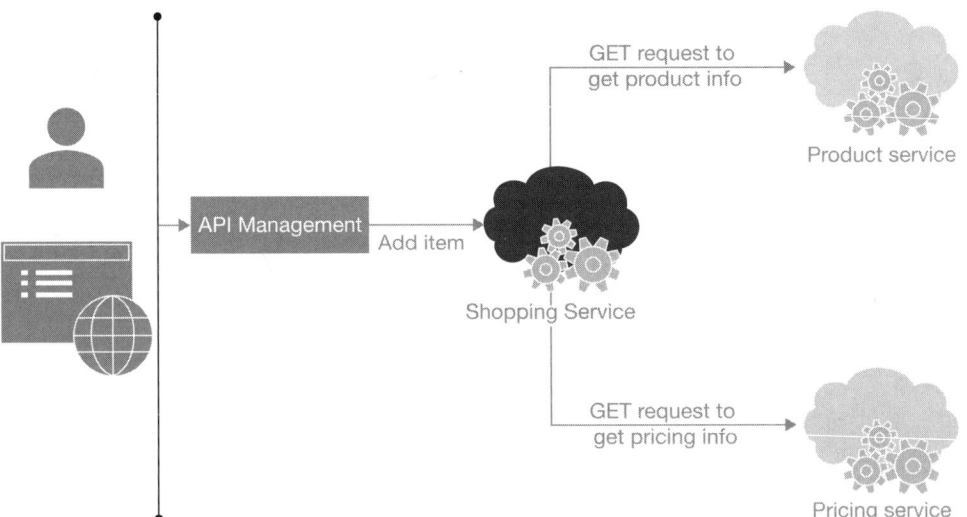

FIGURE 3-8 Using a Direct HTTP call to establish communication between services

Aggregator pattern

An aggregator pattern is another approach to delegating direct HTTP calling from the service itself to an aggregator service, whose responsibility is to make HTTP calls to other services, aggregate the results, and make a single HTTP call to the target service.

In the same shopping service scenario used in Figure 3-8, the request first comes to the aggregator service that makes the request to the product and pricing services. Once the responses are received, it aggregates the results and makes an HTTP POST request to the shopping service. The aggregator service is resilient in nature and should have some mechanism to retry calls if the product and pricing services fail. See Figure 3-9.

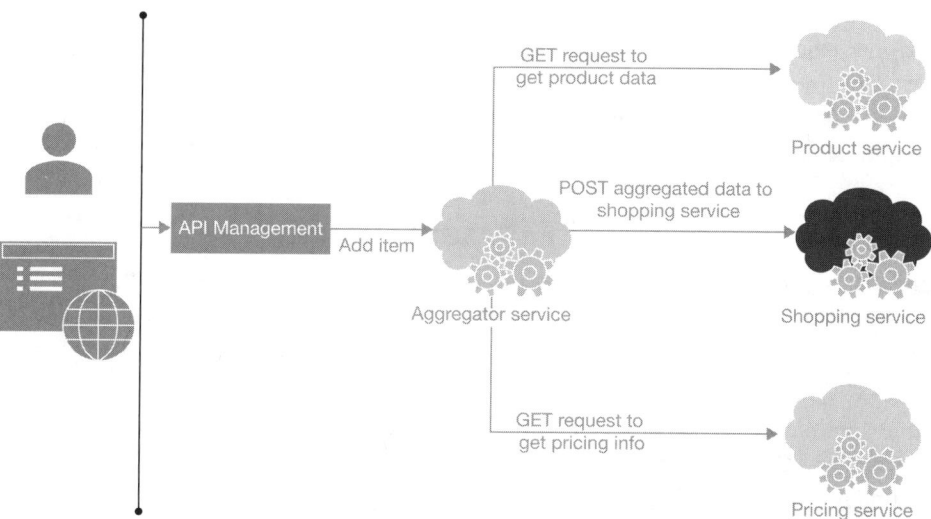

FIGURE 3-9 The aggregator pattern consumes the product and pricing services and sends the aggregated data to the shopping service

The shopping service should only be invoked once the required data has been aggregated by the aggregator service. Alternatively, you can also build some logic into the aggregator service to persist the request and execute it through a background service to overcome failures.

We can also use Azure APIM to build an aggregator service. With Azure APIM, we can read the incoming request and make HTTP requests to other services and aggregate the data. This is possible using policies in APIM. In this case, the architecture looks like Figure 3-10.

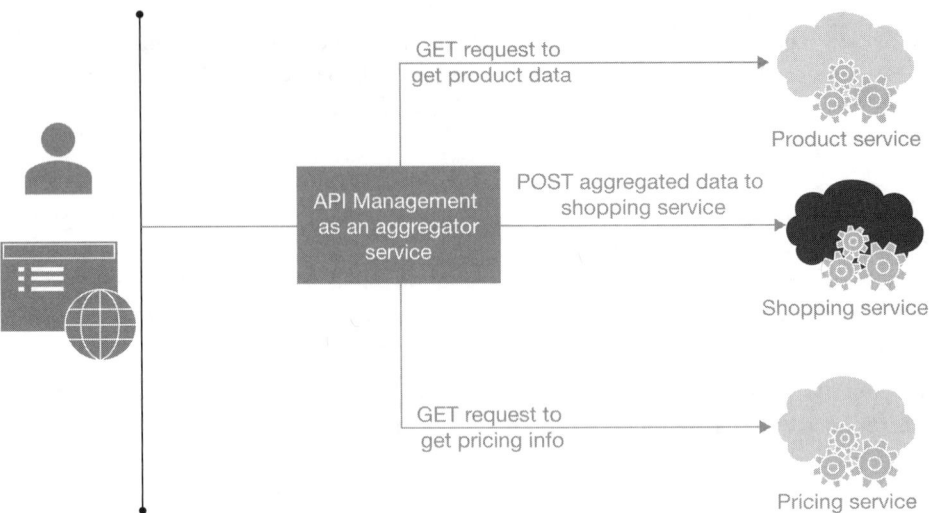

FIGURE 3-10 Using API Management to implement an aggregator pattern

As shown in Figure 3-10, once the Azure APIM receives the request from the website, it makes HTTP Get requests to the product and pricing services, aggregates the results, and then makes a final HTTP POST request to the shopping service to save the shopping transaction to the shopping service database.

Command query responsibility segregation

In the previous approaches, which both used direct HTTP calling, an aggregator is based on an HTTP request. However, sometimes this is not applicable when performance is a primary factor. The service can experience network latency and receive request timeouts each time a call is made to the aggregator or shopping service.

Another approach to implementing the same shopping service scenario is to implement a CQRS (Command Query Responsibility Segregation) pattern. A CQRS pattern separates the read and update operations of a data store.

Figure 3-11 shows a shopping service that has its own read model, which is being synced by the product and pricing data stores. This syncing can be done using a message broker technique. For example, a message broker technique is useful when any insert, update, or delete operation performed on the product or pricing data store triggers an event and pushes a message to a queue that eventually synchronizes the read model part of the shopping service. When a request comes to the shopping service, instead of making calls to the product and pricing services, the shopping service reads the local copy of the data that holds information about the product and pricing data and either saves the transaction or returns a valid response.

The read model of the shopping service can have data in a flat schema. Because we just need to pull out the information to save the shopping transaction, we don't necessarily need to normalize the read model and create multiple tables for both the product and pricing information.

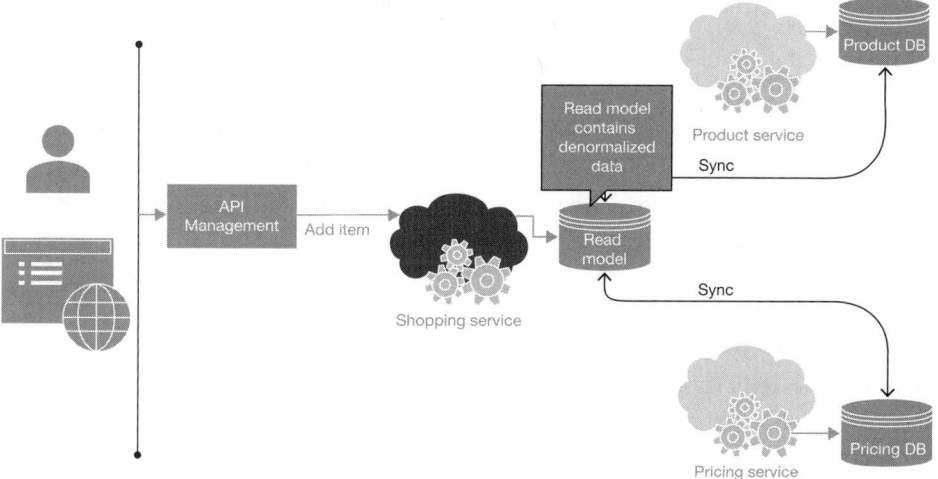

FIGURE 3-11 CQRS pattern

We implemented the CQRS pattern in the OAS. The scenario starts when a bidder bids on active auctions. On every bid, we save that bid information to the bid service database, which contains the list of all bids made on a specific auction. When the auction expires, the system needs to know the highest bidder in order to award the auction. To accomplish this, we can make a direct HTTP call to the bid service to read the last bid information made on that auction. However, with CQRS, we can synchronize a portion of the bid information that holds the bidder ID, bid amount, and so on to the auction database itself. When a bid is made, we publish a message to the message broker. That message contains information about the bid and updates the auction table by making a call to the auction service to update the last bid information to the auction table.

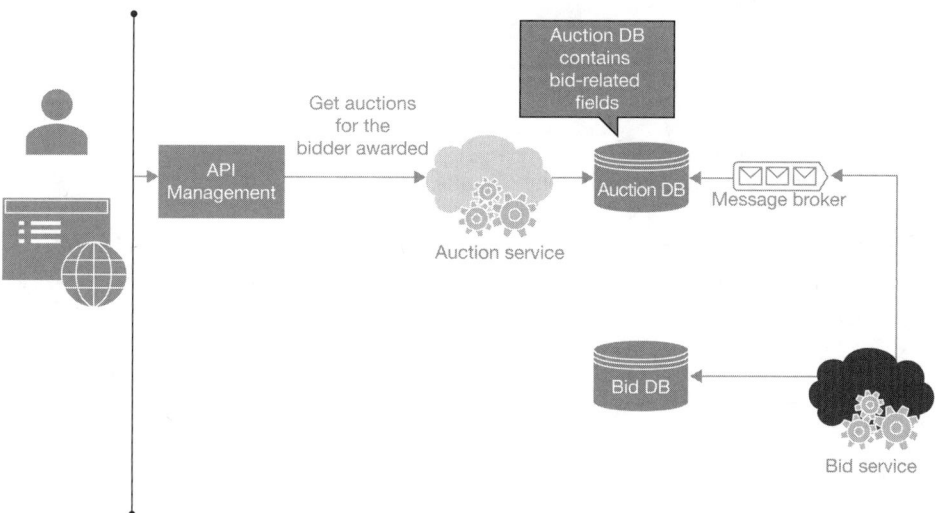

FIGURE 3-12 Implementation of CQRS in the OAS

Saga pattern

In microservices architecture, one of the primary challenges is to maintain consistencies across databases. Distributed transactions are not easily possible with microservices. If a single transaction spans multiple services, keeping consistencies across databases is not an easy task.

This is where the saga pattern comes into practice. Saga patterns enforce data consistency across services. With a saga pattern, each service publishes an event after performing its own transaction. Every other service in the transaction is subscribed to the previous transaction event. Once the event has been published, the subscribers perform their transactions and publish other events that trigger the next service in the transaction. If failures occur, the failed event is generated by any service that is used to roll back the previous transactions.

Figure 3-13 shows a single transaction that spans the order, payment, and notification services. The transaction starts when the order request is submitted. The order request contains the order and payment details. When the order is created, the order service saves the information to the order data store and publishes the order-created event. The order notifies the payment service to perform the payment transaction and to publish a payment-processed event to notify the notification service to send an email notification. If the payment service fails, a new failed event is invoked, which notifies the order service to roll back the order transaction and delete the entry from the order database.

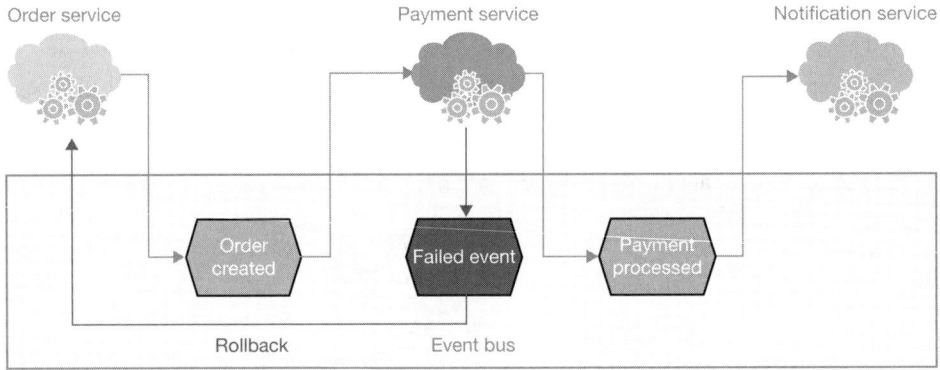

FIGURE 3-13   The saga pattern handling a distributed transaction

With a saga pattern, we can maintain data consistency across multiple services without keeping tight dependencies between them. However, a saga pattern increases the complexity in terms of implementation.

Summary

Software architecture designing is one of the main phases of the software development life-cycle. In this chapter, you learned the following topics:

- We began with a brief introduction to cloud-native applications, as well as their tenets and characteristics.
- We discussed the twelve-factor app methodology, which is one of primary methodologies used with microservices applications.
- We discussed the overall architecture of the Online Auction System and the technologies used in it.
- We discussed patterns related to distributed database architecture.

So far, at this stage of the book, we have designed and discussed the microservices architecture. Going forward, we will be working on developing the front-end application and microservices for the backend on various open-source technologies.

Chapter 4

Develop microservices and front-end applications

In this chapter, you will:

- Build the Online Auction System
- Learn the sequential flows for each business subdomain
- Build microservices in Node.JS, JavaSpring Boot API Framework, and .NET Core
- Provision databases on Azure and use them in services
- Develop an application front-end using Angular

The OAS (Online Auction System) is a public-facing, web-based application that allows anyone to register with the system. Choosing the platform for building applications is one of the primary areas to think about. Today, web applications have been shifted from a server-side model to a client-side model. Users want an intuitive experience where there are fewer or no server-side postbacks and request timeouts. Current industry trends show that web-based applications use AJAX requests to make asynchronous postbacks to the server instead of reloading the whole page. Today, web applications are not limited to running on PCs; web apps can run on mobile and other platforms as well.

On the other hand, back-ends have evolved rapidly, too. Not that long ago, the back-end was connected to the front-end. When a user takes action on the front-end app, the respective event is registered on the back-end listening to that event, which processes the event and returns a response. Nowadays, the application back-end has been exposed as services. Instead of having direct references to the business logic, you call respective RESTful services that perform the back-end work and make it loosely coupled from a front end application. A front-end application communicates to the back-end over the HTTP protocol. The service that exposes the back-end understands messages in JSON or XML payloads. The front-end sends messages in JSON, which get serialized into a respective object on the back-end side, and when a response is returned, it is then deserialized to a respective content type. On both ends, some sort of serialization and deserialization techniques are implemented to read and write messages. In the OAS (Online Auction System), we will develop a SPA (Single Page Application) that communicates to microservices, exposing back-end functionality over HTTP.

Developing microservices

Generally, in big organizations, there are separate teams to build front-end and back-end applications. On the other hand, if team size is small, a single team is sometimes responsible to build the whole application. When building microservices, the split between teams should be directly proportional to the bounded context of a business use case. As a rule, one team can work on one or more bounded context(s), but multiple teams cannot work on a single bounded context.

One of the main benefits of microservices architecture is the ability to use multiple technologies, depending on your needs and specifications. Therefore, for the Online Auction System (OAS), we will demonstrate this value by using polyglot technologies for microservices.

We will be developing these back-end services in this section:

- **Auction service**   A separate service and a MySQL database holds auction information.
- **Place bid service**   A separate service with a Azure Cosmos DB API for MongoDB database holds bid information.
- **Payment service**   A separate service and Azure SQL database holds payment information.

Figure 4-1 shows the auction, bid, and payment services that are built on Node.JS, JavaSpring Boot API, and .NET Core and that use databases such as MySQL, Azure Cosmos DB for MongoDB API, and Azure SQL to store domain-specific information.

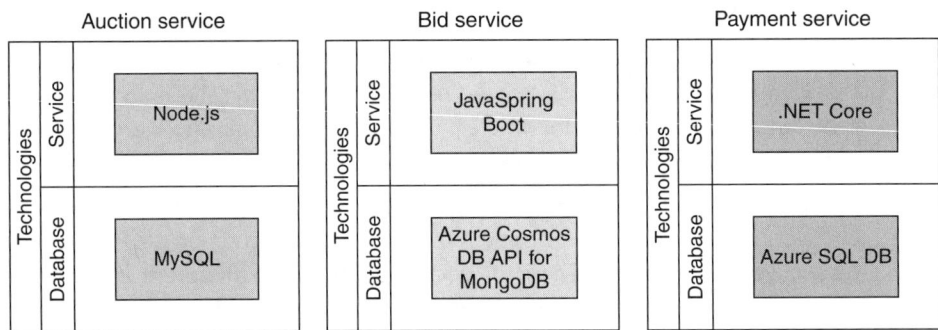

FIGURE 4-1   Technologies used to develop microservices

Developing the auction service

In this section, we will develop an auction service that provides certain methods to create, update, and get auction information. Figure 4-2 shows the sequential flow of an auction subdomain.

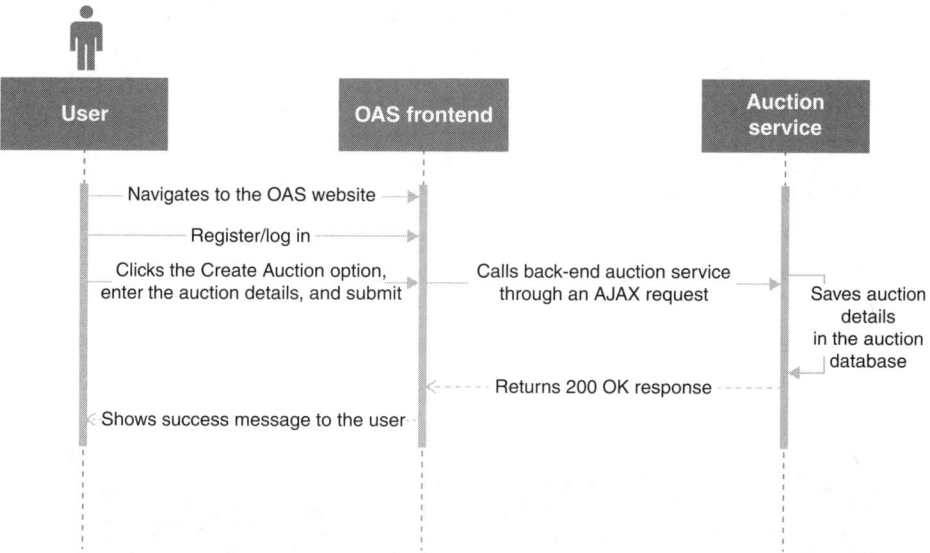

FIGURE 4-2 The Create Auction flow of the auction service

Figure 4-2 shows the flow of an auction. The user needs to log in to the system to create or bid on auctions. Once the user is logged in to the system, he or she can proceed to the **Create Auction** page where the user can specify the auction name, description, hours for which the auction will remain active, and an auction item image. Once the mandatory auction fields are defined, the user can click the **Save** button to submit the auction by making a call to the auction service.

These prerequisites must be installed on a development machine:

- **Node.js** Download from *https://nodejs.org/en/download/*
- **VS Code** Download from *https://code.visualstudio.com/download*
- **MySQL Workbench** Download from *https://www.mysql.com/products/workbench/*

Provision a MySQL database in Azure

The auction service uses MySQL database to hold auction information. MySQL on Azure is a managed database in the cloud that can be provisioned from Azure portal or Azure CLI or by using ARM (Azure Resource Manager) or Terraform templates. In this section, we will create a Terraform template to provision a MySQL resource on Azure.

> **NOTE** Terraform is an open-source infrastructure-as-code (IaC) tool created by HashiCorp. It enables the user to create templates using the HashiCorp configuration language or JSON and to provision it to various cloud platforms using commands.

First, you must configure Terraform. To install it on Windows, you need to have Chocolatey installed and then run the following command from PowerShell:

```
Choco install terraform
```

> **NOTE** **Setting up Terraform**
>
> To set up Terraform on your development machine, see
> https://learn.hashicorp.com/terraform/getting-started/install.html#overview.

There are many ways to write Terraform templates. However, we will use a VS Code Terraform extension to create a Terraform template. You can download the Azure Terraform extension from Microsoft from the VS Code extensions palette, as shown in Figure 4-3.

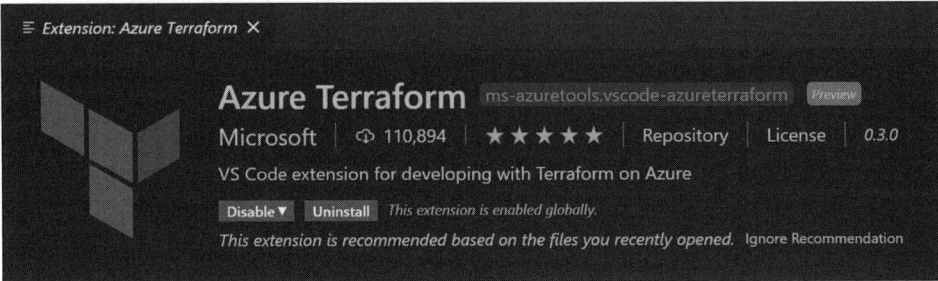

FIGURE 4-3   The Azure Terraform extension in VS Code

> **NOTE** **Extensions palette shortcut key**
>
> You can open the extensions palette by pressing Ctrl+Shift+X.

Once the extension is installed, reload VS Code, create a new Terraform file, and name it MySQL.tf. Use the script shown in Listing 4-1 to provision a MySQL resource with Terraform.

LISTING 4-1   Provisioning a MySQL resource with Terraform

```
provider "azurerm" {
  features {}
}
resource "azurerm_mysql_server" "resource-values" {
  name                = "mysqloas"
  location            = "West Europe"
  resource_group_name = "OSS"

  administrator_login          = "mysqladmin"
  administrator_login_password = "P@ssw0rd!@#"
```

```
sku_name    = "B_Gen5_2"
storage_mb = 5120
version     = "5.7"

auto_grow_enabled                    = true
backup_retention_days                = 7
ssl_enforcement_enabled              = false
}
```

In the script shown in Listing 4-1, we first initialized the provider as azurerm. This tells the Terraform that we are creating the resource in Azure. Then we configured the resource and declared the type as azurerm_mysql_server. Inside the braces, we then mentioned the properties, such as resource group name, location, administrator login and password, and other database-related values.

To run the script, open the command palette in VS Code and type the Azure Terraform: Init command, as shown in Figure 4-4.

FIGURE 4-4 Initializing the cloud shell for Azure Terraform

This command prompts you to sign in to the Azure portal and open a cloud shell in VS Code. Once the cloud shell opens, run terraform init, either from a cloud shell or from the command palette.

You will see the message in Figure 4-5 in the cloud shell once the terraform init command has been successfully executed.

```
Terraform has been successfully initialized!

You may now begin working with Terraform. Try running "terraform plan" to see
any changes that are required for your infrastructure. All Terraform commands
should now work.

If you ever set or change modules or backend configuration for Terraform,
rerun this command to reinitialize your working directory. If you forget, other
commands will detect it and remind you to do so if necessary.
ovais@Azure:~/clouddrive/terraforms$ []
```

FIGURE 4-5 The cloud shell after running the Azure terraform init command

Next, run the terraform plan command to validate the script. You can run it by typing the command in the cloud shell or the command palette. Finally, call the terraform apply command to make the changes and provision resources in the cloud. See Figure 4-6.

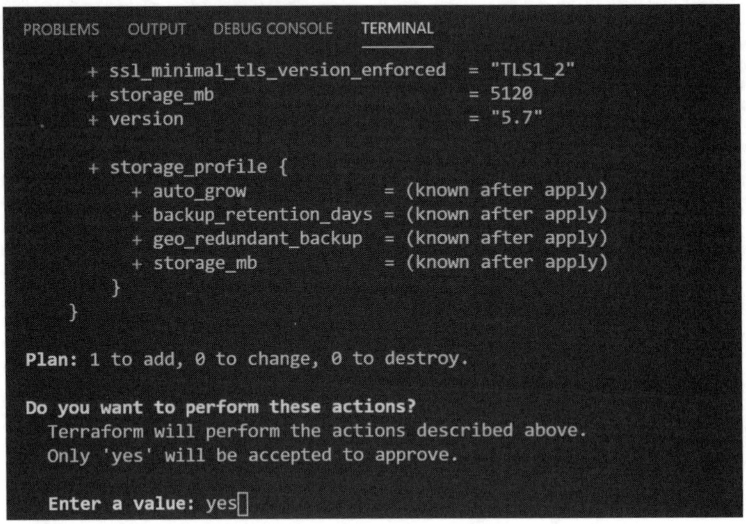

```
PROBLEMS    OUTPUT   DEBUG CONSOLE    TERMINAL

    + ssl_minimal_tls_version_enforced   = "TLS1_2"
    + storage_mb                         = 5120
    + version                            = "5.7"

    + storage_profile {
        + auto_grow              = (known after apply)
        + backup_retention_days  = (known after apply)
        + geo_redundant_backup   = (known after apply)
        + storage_mb             = (known after apply)
      }

  }

Plan: 1 to add, 0 to change, 0 to destroy.

Do you want to perform these actions?
  Terraform will perform the actions described above.
  Only 'yes' will be accepted to approve.

  Enter a value: yes▯
```

FIGURE 4-6 Applying the deployment and provisioning of Azure resources using Terraform

Once the `terraform apply` command is successfully executed, you can verify the resource creation by logging in to the Azure portal.

Create a database

The Auction database contains a single table to hold auction-related information. When designing databases for microservices, normalization is not necessary for it to be applied in relational databases. Because the database targets one business domain, it contains tables that only belong to that business domain.

In the auction table, we will be keeping some information related to the bid service as well. And this is intentionally designed in a way to implement sort of a CQRS (Command Query Responsibility Segregation) pattern. That means when the bid is recorded in the bid database, the last bid information is updated in the auction database. Table 4-1 shows the schema of the auction table.

TABLE 4-1 Auction table schema

Column Name	Type	Description
idAuction	int(11)	Primary key, unique auto numerical value
Name	varchar(45)	Name of the auction
Description	varchar(300)	Description of the auction
StartingPrice	decimal(6,0)	Starting price set by the user when creating the auction
AuctionDate	Date	Date and time auction was created
Status	int(11)	Status can be `Active` or `Completed`
Image	mediumtext	Holds the binary value of the image

Column Name	Type	Description
ActiveInHours	int(11)	Number of hours the auction remains active. The number of hours added in `AuctionDate` to determine the cutoff time
BidPrice	decimal(6,0)	Last bid price value
UserId	varchar(250)	User who created the auction
IsActive	tinyint(4)	Value can be `Active` or `InActive`
UserName	varchar(300)	Name of the customer who placed the last bid
IsPaymentMade	tinyint(4)	Boolean value indicating if payment has been made
BidUser	varchar(250)	Bidder ID
BidId	varchar(250)	Record ID

The fields mentioned in Table 4-1 are self-explanatory. However, fields such as `BidId`, `BidPrice`, `BidUser`, and `UserName` hold bid specific information that is updated when a bid is made on an active auction.

To create a similar schema in the newly created MySQL database, you can download the MySQL workbench tool and connect to the database by providing the `hostname`, `username`, and `password`.

> **NOTE** **Add the client IP**
>
> Make sure to add the client IP from the connection security tab in the Azure portal for MySQL database. Otherwise, you won't be able to connect to the database.

Listing 4-2 shows the script to create an auction table in the MySQL database.

LISTING 4-2   Creating a new auction table

```
CREATE TABLE `auction` (
  `idauction` int(11) NOT NULL AUTO_INCREMENT,
  `name` varchar(45) DEFAULT NULL,
  `description` varchar(300) DEFAULT NULL,
  `startingPrice` decimal(6,0) DEFAULT NULL,
  `auctionDate` date DEFAULT NULL,
  `status` int(11) DEFAULT NULL,
  `image` mediumtext,
  `activeInHours` int(11) DEFAULT NULL,
  `bidPrice` decimal(6,0) DEFAULT NULL,
  `userId` varchar(250) DEFAULT NULL,
  `isActive` tinyint(4) DEFAULT NULL,
  `userName` varchar(300) DEFAULT NULL,
  `isPaymentMade` tinyint(4) DEFAULT NULL,
```

```
  `bidUser` varchar(250) DEFAULT NULL,
  `bidId` varchar(250) DEFAULT NULL,
  PRIMARY KEY (`idauction`)
) ENGINE=InnoDB AUTO_INCREMENT=54 DEFAULT CHARSET=latin1;
```

Create a new database, name it `auctionservicedb`, and then run the above script to create the auction table in the MySQL database using the MySQL Workbench tool.

Create Auction Service in Node.JS

Node.js is one of the core platforms for developing applications in the JavaScript and Type-Script languages. There are lot of frameworks available to develop different kinds of applications. For example, to develop APIs, there are frameworks such as Express.JS and Molecular; for desktop applications, there are frameworks such as Electron and Meteor; and for web applications, there are frameworks like Angular and React.

To develop the auction service, we chose Express.JS because it's good for developing RESTful APIs and provides a simple an easy way to write API methods. First, create a new folder named `auctionservice` and open VS Code. To create the Express.JS project, you can run the following command from the VS Code terminal window or from the command prompt.

```
npm install -g express-generator
```

The `npm` is the node package manager used to install node modules. The express-generator can be installed as a node module by running the above command. The -g flag is used to globally install the node module in the system. Once the Express generator module has installed successfully, the Express project can be created by running the following command:

```
express
```

The above command creates an `express` project that contains the following folders and the `app.js` file that is used as an entry point to trigger the Express app to listen for HTTP requests.

- **Public folder**   Contains static files, images, and scripts
- **Routes folder**   Contains APIs
- **Views folder**   Contains user interfaces

For the auction service, we will be developing API methods inside the Routes folder.

Write API Methods for the auction service

The auction service is used to manage auction items. We need to expose certain APIs that can be consumed by the web application to read auctions, to get auctions by ID, and create auctions. Table 4-2 shows the list of methods we will be developing in this section.

TABLE 4-2 Auction service API methods

API Method	HTTP Verb	Signature	Description
Get Auctions	GET	/auctions	Returns all the active auctions
Get Auctions by Bid User	GET	/auctionsbyBidderId/:bidUserId	Returns the active auctions based on the user who bid on the auction
Get Auctions based on User ID	GET	/auctionsByUserId/:userId	Returns the list of auctions created by the user
Get Auctions based on Auction ID	GET	/auctionById:id	Returns the list of auctions by auction ID
Create Auction	POST	/auctions	Creates a new auction
Update Auction for Bid	PUT	/updateAuctionForBid	Updates the auction table with the last bid information

To connect with MySQL database, we need to first install the `mysql` package using the `npm` command. Execute the following command to install `mysql`.

```
npm install -g mysql
```

Next, create a `config.js` file under the `public/data` folder and add the script shown in Listing 4-3:

LISTING 4-3 Connecting with a MySQL database

```
const mysql = require('mysql');
// Set database connection credentials
const config = {
 host: 'mysqloas.mysql.database.azure.com',
 user: 'user@mysqloas',
 password: 'P@ssw0rd!@#',
 database: 'auctionservicedb',
};
// Create a MySQL pool
const pool = mysql.createPool(config);
// Export the pool
module.exports = pool;
```

In the above script, initialize the `mysql` argument first by using the `require` argument. Then create a config file that holds the connection information, such as host, user, password, and database. Finally, call the `createPool` method to create the pool object by passing the configuration object. The `module.exports` command is used to export the object to be used by other programs.

Now, we can use the `pool` object in our `index.js` file and add methods for different scenarios. To get the list of active auctions, we can run the inline query as shown in Listing 4-4:

LISTING 4-4 Retrieving an active auction list

```
// Display all Auctions
router.get('/auctions', (request, response) => {
```

```
client.trackTrace("Loading all auctions");
pool.query('SELECT idAuction, Name, Description, StartingPrice, AuctionDate,
Status, Image, userName,  DATEDIFF(date_add(auctiondate, interval activein-
hours hour), curdate()) * 24  as ActiveInHours, BidPrice  from auctionservicedb.
auction where DATEDIFF(date_add(auctiondate, interval activeinhours hour),
curdate()) >=0  and IsActive=1', (error, result) => {
    if (error) throw error;
    response.send(result);
  });
});
```

To get the list of auctions based on bidder ID, we can create a method as shown in Listing 4-5:

LISTING 4-5 Auctions based on bidder ID

```
router.get('/auctionsbyBidderId/:userId', (request, response) => {
  client.trackTrace("Loading all auctions by user id");
  const userId = request.params.userId;
  pool.query("SELECT * from auction where IsActive=1 and bidUser ='"+userId+"'",
  (error, result) => {
      if (error) throw error;
      response.send(result);
    });
});
```

You can refer to the code repository for rest of the methods. (See the Introduction at the beginning of this book for details about accessing and downloading the code for this book.)

Developing the bid service

In this section, we will create a bid service that provides methods to create, update, and get bid information. Figure 4-7 shows a sequential flow for creating bids.

The process flow starts when the user navigates to the OAS and clicks the **Active Auctions** option. The Active Auctions window shows the list of active auctions. The user selects the appropriate auction that opens another page and shows the auction details. The user can specify the bid amount while placing bids or making offers. Figure 4-8 shows the bid screen where a user can place a bid. From there, users can select the respective auction to submit the bid. Figure 4-9 shows the **Make A Bid** page where a user can see an auction's details, as well as previous bids made by other users. From here, the user can specify the bid amount and submit the bid.

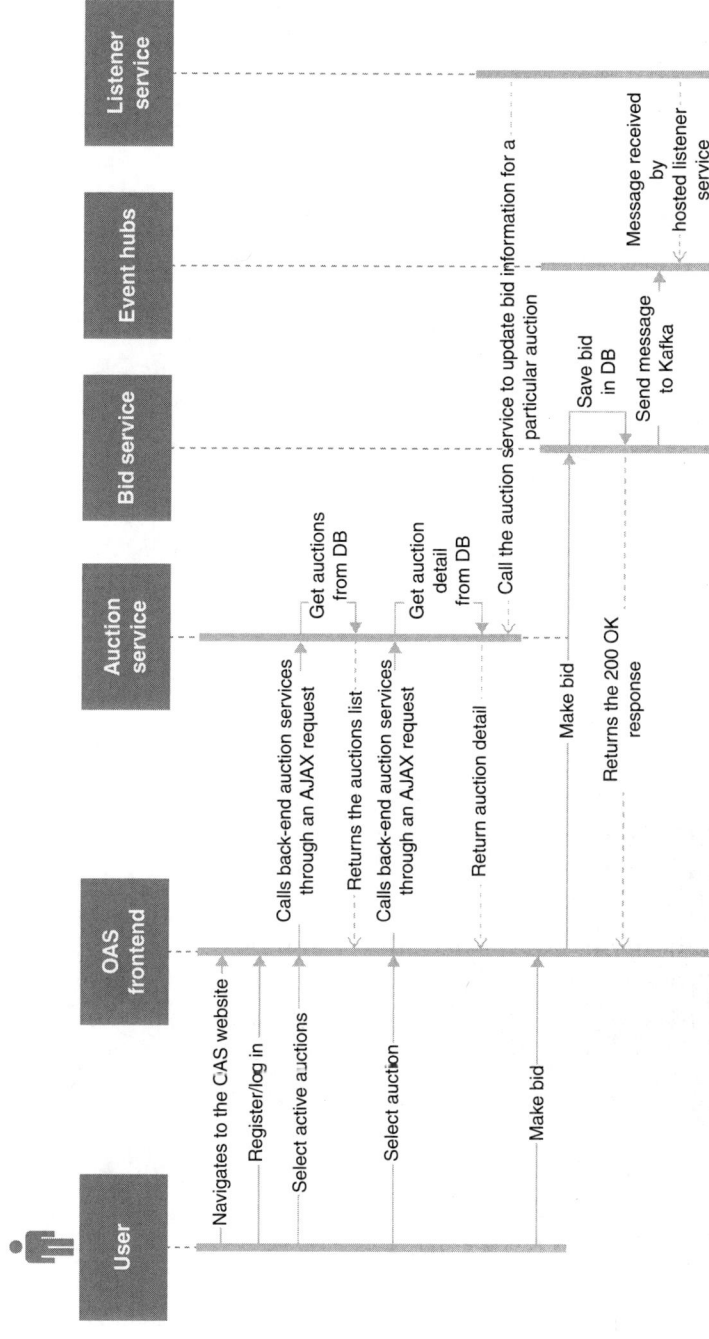

FIGURE 4-7 The complete sequential flow of a bid being made

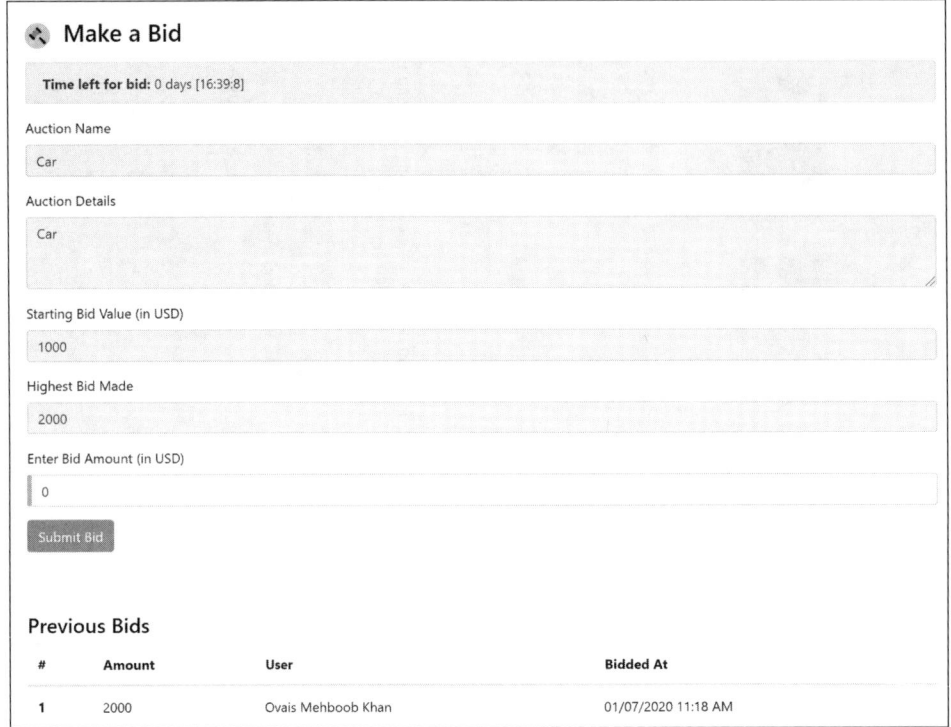

FIGURE 4-8 Active auctions

FIGURE 4-9 Make A Bid page

NOTE **Prerequisites**

To develop the bid service, the following items must be installed on the development machine:

- Java 1.8 *https://www.azul.com/downloads/azure-only/zulu*
- VS Code *https://code.visualstudio.com/download*

Provision Cosmos DB in Azure

The bid service uses Cosmos DB for the MongoDB API database to hold bid-related information. Azure Cosmos DB is a native cloud database that is designed to handle cloud workloads—unlike other databases, which are considered to be immigrants to cloud platforms. The Azure

Cosmos DB is a managed service that is globally distributed and supports multi-model databases. With Azure Cosmos DB, you can create both SQL and NoSQL databases. Azure Cosmos DB supports the following APIs.

- Azure Cosmos DB SQL API
- Azure Cosmos DB API for MongoDB
- Azure Cosmos DB Cassandra API
- Azure Cosmos DB Gremlin API
- Azure Cosmos DB Table API

We can choose any API on which to build the database, and you can use the same native SDK to connect to it. Azure Cosmos DB supports elastically scaling to various Azure regions worldwide. You can elastically scale throughput and storage and benefit from swift data access using any of the APIs mentioned above.

For the bid service, we will keep the bid information in a NoSQL database and thus, we used Azure Cosmos DB's MongoDB API. We chose the NoSQL database because of the high frequency of bids being made on active auctions and because NoSQL databases are optimized for insertions.

To provision Azure Cosmos DB in Azure, we use the same Terraform extension in VS Code. Make sure the Terraform extension is already installed in VS Code. (If you haven't done this yet, see the previous section.)

We will create a new Terraform file and name it `AzureCosmosDB.tf`. Listing 4-6 shows the script to provision Azure Cosmos DB for a Mongo API resource with Terraform:

LISTING 4-6 Provisioning an Azure Cosmos DB for a Mongo API resource with Terraform

```
resource "azurerm_cosmosdb_account" "cosmos-db" {
  name                = "mongodboas"
  location            = "westeurope"
  resource_group_name = "OSS"
  offer_type          = "Standard"
  kind                = "MongoDB"
  consistency_policy {
    consistency_level       = "BoundedStaleness"
    max_interval_in_seconds = 10
    max_staleness_prefix    = 200
  }
  geo_location {
    location          = "westeurope"
    failover_priority = 0
  }
}
```

In Listing 4-6, the code is broken down like so:

- The `name` argument contains the actual name of the Cosmos DB resource you are provisioning, and the `location` argument is the Azure region where the resource will be provisioned.

- The `resource_group_name` argument states the name of the resource group, which is the same resource group we specified for MySQL server database.

- The `offer_type` argument denotes the plan. `Kind` should be set to MongoDB in order to create Azure Cosmos DB's API for MongoDB.

- The `consistent_policy` argument holds information specific to the consistency level and the max interval (in seconds) to represent the amount of staleness (in seconds) tolerated. The staleness prefix indicates the number of tolerated stale requests.

- Finally, the `geo_location` argument holds information specific to the location we need to initial provision this, and the `failover_priority` argument is set to 0, which indicates a write region. The failover priority equals the total number of regions minus one.

To run the above script, you need to first run the `terraform init` command from VS Code, which opens the cloud shell in a VS Code terminal window. Next, you need to run `terraform plan` and then `terraform apply` as shown earlier in this chapter to create Azure Cosmos DB resource in Azure. See "Provision a MySQL database in Azure."

Database schema

The bid service database schema is shown in Table 4-3.

TABLE 4-3 Bid service database schema

Column Name	Type	Description
bidId	String	The unique ID of a bid
auctionId	String	An auction's ID
bidAmount	Number	The bid amount
userId	String	Bidder's user ID
bidDate	Date	Date when the bid was made

Create a bid service in the JavaSpring Boot framework

The bid service will be developed using the JavaSpring Boot framework, which is one of the most popular frameworks used for developing RESTful APIs in Java. You can easily create enterprise-grade APIs and accommodate all those scenarios that are essentials to be considered such as applying annotations and integrating with the Spring ecosystem that includes Spring JDBC, Spring ORM, and so on.

The JavaSpring Boot framework offers two flavors for creating APIs: Maven and Gradle. For the bid service, we will provision an application using Maven. First, create a new VS Code workspace/folder where the application will be created. VS Code provides an extension to provision the Spring Boot API.

Install the Java Extension Pack extension by going to the extension's palette from VS Code, as shown in Figure 4-10.

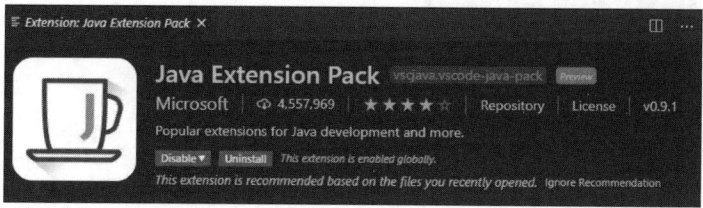

FIGURE 4-10   VS Code Java Extension Pack

Once the extension is installed, open the command palette by pressing Ctrl+Shift+P in VS Code, typing `spring`, as shown in Figure 4-11, and selecting the `Spring Initializr: Generate a Maven Project` option.

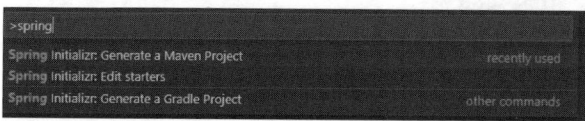

FIGURE 4-11   The above figure shows the way of creating JavaSpring Boot application using the Java Extension Pack installed in the previous step.

Once the **Maven** option is selected, it will take you through some wizard steps:

1.   Under **Specify Project Language**, choose **Java**, as shown in Figure 4-12.

FIGURE 4-12   Select Java

2.   For the I**nput Group Id For Your Project**, enter **com.onlineauctionweb,** as shown in Figure 4-13.

FIGURE 4-13   Group Id

3.   For the **Input Artifact ID For Your Project**, enter **bidservice**, as shown in Figure 4-14.

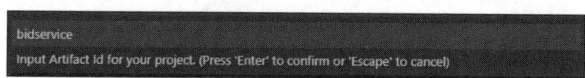

FIGURE 4-14   The Artifact ID is bidservice

4. Choose the JavaSpring Boot Application version. For our purposes, we will select **2.3.1**.

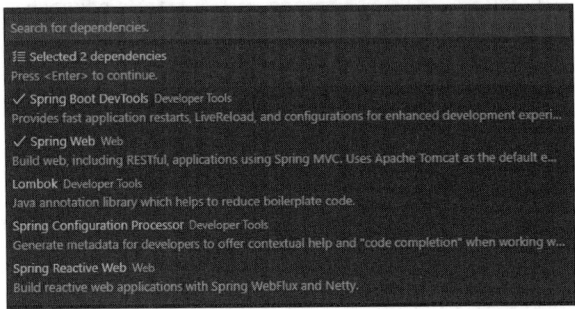

FIGURE 4-15 Choose the JavaSpring Boot application version

5. Select **Spring Boot DevTools** and **Spring Web** as dependencies, as shown in Figure 4-16.

FIGURE 4-16 Select the dependencies used in the JavaSpring Boot Application

6. Choose the folder in which you will create the project files. You can choose the **bidservice** folder, where all the project files will be created.

7. Once the project is created, another instance of the VS Code is opened where the project is opened.

API methods in the bid service

The bid service is used to create bids. Table 4-4 shows the list of methods we will be adding in the bid service.

TABLE 4-4 Bid service methods

API Method	HTTP Verb	Signature	Description
Get Bid by Auction ID	GET	/bid	Returns bids based on Auction ID
Create Bid for Auction	POST	/bid	Creates a bid for the active auction

Create a bid model

Now that the basic project has been scaffolded, the first thing we need to do is create a bid model. This model contains some properties related to the bid table and is used for serializing/deserializing the JSON payload into the object.

Create a new folder named Models inside the /bidservice folder and add a new file named BidDetail.java. Listing 4-7 shows the code snippet of the BidDetail class:

LISTING 4-7   Bid detail entity

```
public class BidDetail{
    public String bidID;
    public String amount;
    public String customer;
    public String customerName;
    public String bidAt;
}
```

The BidDetail contains properties such as bidID, amount, customer, customerName, and bidAt.

Create a bid controller

Add a bid controller that will expose some methods to create and get bids from the Mongo DB database, and add a new class inside the /bidservice folder and name it BidController.

Get bids using the auctionID method

Add a method to return the bids by auctionID. Add a new method named GetBidByAuctionID that takes auctionID as a parameter and returns the list of auctions. Listing 4-8 shows the code snippet for the GetBidByAuctionID method.

LISTING 4-8   Get Bid by Auction ID method

```
        @CrossOrigin(origins = "*", allowedHeaders = "*")
        @RequestMapping(value = "/bid", method = RequestMethod.GET)
        @ResponseBody
        public ArrayList GetBidByAuctionID(String auctionID)
        {
            ArrayList lst=new ArrayList();
             telemetryClient.trackTrace("Getting Bid by Auction ID");
            try {
                MongoClient mongoClient = new MongoClient(new MongoClientURI(
                        "mongodb://mongodbtfx-vSaxMiCzTRAly2sL4H8IqpkidImxHK36G3H8oIKb9q
                        0fCQbjfSuMOHqC8pwRzpSLtOWbrfITjzsOe2FYDidWHw==@mongodbtfx.docu-
                        ments.azure.com:10255/?ssl=true&replicaSet=globaldb"));
                DB db=  mongoClient.getDB("biddb");
                DBCollection coll=db.getCollection("bids");
                BasicDBObject doc = new BasicDBObject();
                doc.put("auctionId", auctionID);
                DBCursor cursor= coll.find(doc);
                Integer counter=1;
                while(cursor.hasNext()) {
```

```
                    BasicDBObject obj = (BasicDBObject) cursor.next();
                    BidDetail bidDetail=new BidDetail();
                    bidDetail.bidID = counter.toString();
                    bidDetail.amount = obj.getString("bidAmount");
                    bidDetail.customer = obj.getString("userId");
                    bidDetail.customerName= obj.getString("userName");
                    bidDetail.bidAt=obj.getString("bidDate");
                    lst.add(bidDetail);
                    counter++;
                }
            } catch (UnknownHostException e) {
                // TODO Auto-generated catch block
                e.printStackTrace();
            }
        return lst;
    }
```

In the method shown in Listing 4-8, we are first initializing the `ArrayList` instance and then initializing the `MongoClient` object and passing the connection string. The connection string can be obtained from the Azure Portal, by navigating to the Cosmos DB resource and opening the **Connection Strings** tab. You need to replace the connection string to connect with the Cosmos DB resource. Next, initialize the DB object and mention the exact name of DB, which in our case is `biddb`. The `DBCollection` is used to read the collection by specifying the collection name.

To find the bids based on Auction ID, use the `BasicDBObject` class and specify the key/value pair where the key refers to the column name in the bids collection and the value represents the parameter value passed in the method.

Also, you need to initialize the counter to start at 1 and then execute a loop to read all the bids for the auction. Finally, return the list of bids as a response. To use MongoDB API, you also need to modify the `pom.xml` file and add the MongoDB dependency as shown in Listing 4-9:

LISTING 4-9 Adding MongoDB library reference

```
<dependency>
        <groupId>org.mongodb</groupId>
        <artifactId>mongo-java-driver</artifactId>
        <version>2.12.4</version>
</dependency>
```

See the code repository mentioned in this book's Introduction for a complete list of dependencies needed to build the application.

Create the bid method

Add a new method to create bid and name it `CreateBid`. The method takes four parameters: `bidAmount`, `auctionID`, `userID`, and `userName`. Listing 4-10 shows the complete `CreateBid` method code snippet:

LISTING 4-10 Create Bid method

```
@CrossOrigin(origins = "*", allowedHeaders = "*")
  @RequestMapping(value = "/bid", method = RequestMethod.POST)
  @ResponseBody
  public void CreateBid(String bidAmount, String userID, String auctionID, String
  userName) throws Exception {
      try {
          telemetryClient.trackTrace("Creating bid");
          MongoClient mongoClient = new MongoClient(new MongoClientURI(
                  "mongodb://mongodbtfx:xSaxMiCzTRAly2sb4H8IqpkidTmxHK36G3H8oIKb9qOfC
                  QbjfSuMOHqC8pwRzpSLtOWbrfITjzsOe2FYDidWHw==@mongodbtfx.documents.
                  azure.com:10255/?ssl=true&replicaSet=globaldb"));
          DB db=  mongoClient.getDB("biddb");
          DBCollection coll=db.getCollection("bids");
          SimpleDateFormat dateFormat = new SimpleDateFormat("yyyy/MM/dd HH:mm:ss");
          LocalDateTime now = LocalDateTime.now();

          BasicDBObject doc = new BasicDBObject();
          doc.append("bidid", UUID.randomUUID().toString());
          doc.append("auctionId", auctionID);
          doc.append("bidAmount", bidAmount);
          doc.append("userId", userID);
          doc.append("userName", userName);
          doc.append("bidDate", System.currentTimeMillis());
          coll.insert(doc);
      } catch (UnknownHostException e) {
          e.printStackTrace();
      }
  }
```

The method shown in Listing 4-10 initializes the mongo client object and then opens a cursor to the MongoBD database and collection. It then creates a new BasicDBObject and specifies the document fields. Finally, it calls the insert method to save that document into the bid database.

Developing a payment service

The payment service is used to pay for the items won based on the highest bid. Once the auction time ends, the highest bidder is awarded the auction. The system allows the user to proceed to the payment page and pay the amount due. Because this is a sample case study, we did not use payment gateway; instead, we have created a service that saves the payment transaction information in a separate payment database. Figure 4-17 shows the sequential flow when a payment is made for a winning auction.

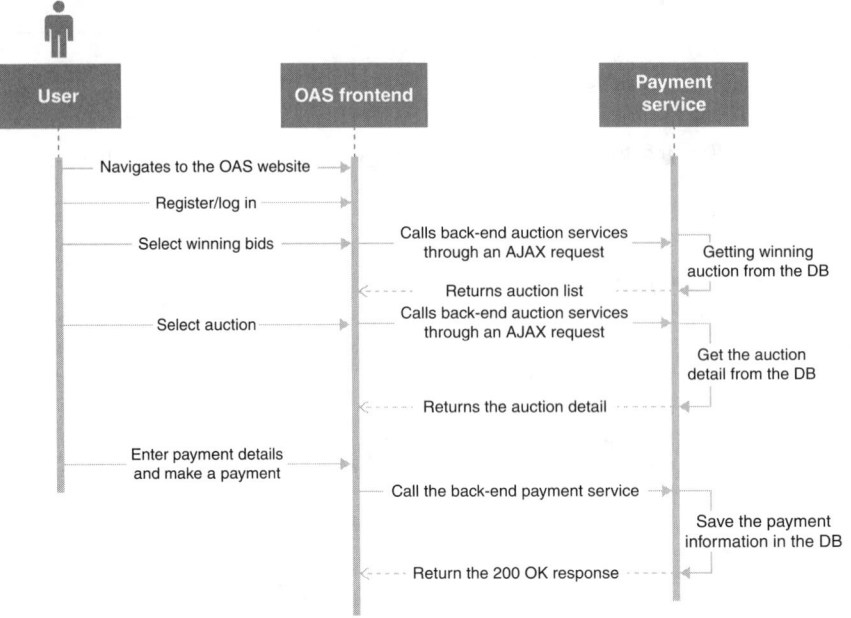

FIGURE 4-17 The sequential flow of making a payment

The payment process flow starts when user navigates to the site and clicks the **Winning Bids** option, which opens a table showing all the auctions that user has won. If the user selects the auction, the auction details are shown. The payment page contains fields to input user payment details and store them in the database.

Figure 4-18 shows the Winning Bids list, where a user can select winning bids and proceed to making a payment.

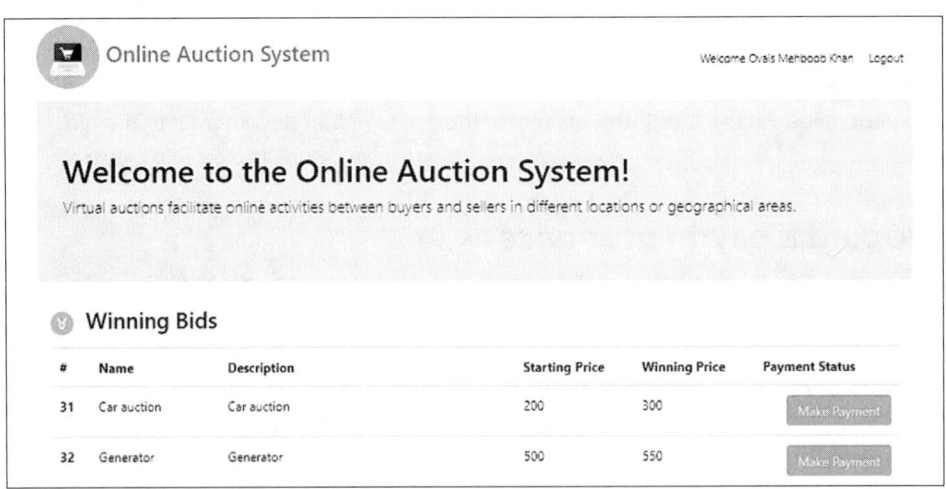

FIGURE 4-18 The Winning Bids screen

Figure 4-19 shows the payment screen from which the user can make payments.

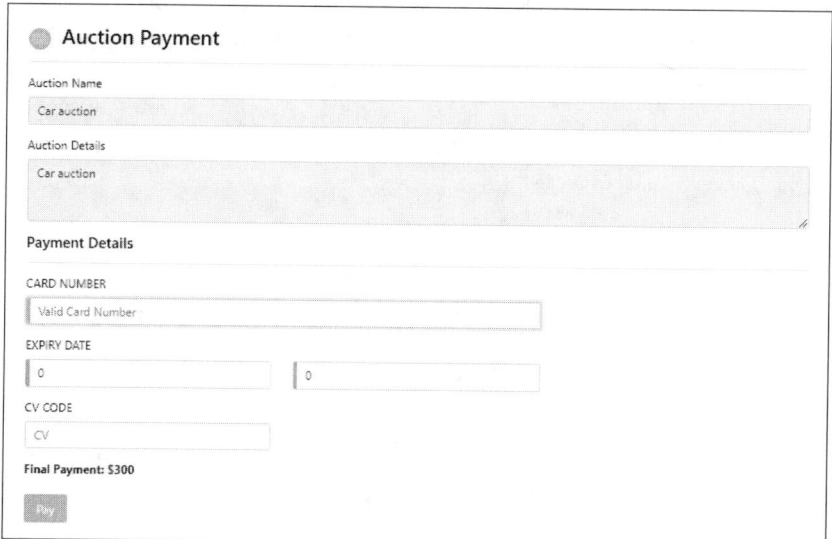

FIGURE 4-19 Auction Payment screen

Create a payment service database

The Payment Service database contains payment transaction information. Therefore, we use a SQL database technology and utilize the Azure SQL database to keep payment information. Azure SQL is a managed cloud database on Azure that runs on a cloud computing platform and provides high availability, scalability, and failover scenarios. There are three kinds of Azure SQL Databases that can be provisioned in Azure as listed below. However, for the bid service, we will be creating an Azure SQL Single Database resource type because there is only one service that will be hosting the database on Azure SQL.

- **Azure SQL Single Database Model**   A managed instance to create a single database.
- **Azure SQL Elastic Database Model**   A shared resource model to host multiple databases that share the resources within an elastic pool. It is a cost-effective model if you want to host multiple databases.
- **Azure SQL Managed Instance**   Similar to an on-premises SQL Server instance, where you can host multiple databases and share resources between them. It provides some instance-level features, such as SQL Agent, Service broker, Linked servers, and many more.

> **MORE INFO**   **Azure SQL Database versus Azure SQL Managed Instance**
>
> For a complete features comparison between Azure SQL Database and an Azure SQL Managed Instance, see *https://docs.microsoft.com/en-us/azure/azure-sql/database/features-comparison.*

Let's provision the Azure SQL Single Database in Azure using Terraform. First, create a new file under the same Terraform workspace in VS Code and name it AzureSQL.tf. Listing 4-11 shows the Terraform script to provision an Azure SQL Server and an SQL database for the payment service.

LISTING 4-11 Provision an Azure SQL Server and SQL database

```
resource "azurerm_sql_server" "oasresource" {
  name                         = "oassqlserver"
  resource_group_name          = "OSS"
  location                     = "West Europe"
  version                      = "12.0"
  administrator_login          = "sqladmin"
  administrator_login_password = "P@ssw0rd!@#"
 }
resource "azurerm_sql_database" "oasresource" {
  name                = "auctionpaymentdb"
  resource_group_name = "OSS"
  location            = "West Europe"
  server_name         = "oassqlserver"
}
```

In Listing 4-11, we first declare the section for Azure SQL Server. Next, we set the name of the SQL Server as oassqlserver, and we set the resource group to OSS, which is the same name we used while provisioning other resources. For the location, we used West Europe. We then set SQL administrator login name and password to log in to the server.

In the next section of the script, we specify the database details, such as the name of the database, resource group, location, and server name. The server name should be the same name you specified while creating the server.

You can run the script shown in Listing 4-11 using Terraform commands such as init, plan, and apply. Once the database is created, you can access it from SQL Server Management Studio or Query editor options from the Azure Portal itself and execute the following script to create a table. In order to access the database from a local machine, you need to add a client IP in Azure Portal at the server level, as shown in Figure 4-20.

FIGURE 4-20 Adding a Client IP from the Azure portal

Because the services will be deployed on Azure, in order to have it accessible from those services, you need to enable the Allow Azure services option from the same tab, as shown in Figure 4-21.

FIGURE 4-21   Allow Azure SQL Server to be accessible by other Azure resources

Now that we have set up the database for the payment service, we will create a payment table using Entity Framework code in the next section.

Create Payment Service in ASP.NET Core

The Payment Service will be developed in ASP.NET Core. The .NET Core is a Microsoft open-source managed framework for developing modern, cloud-based, and high-performance applications that run cross platform. It is one of the best suited frameworks for developing microservices because it's more modular. The developer can build the applications using a set of libraries.

In comparison to the traditional use of ASP.NET Web Forms or MVC in .NET Framework, the application is tightly dependent on the System.Web assembly. System.Web assembly contains lots of packages, and the assembly size is around 5 MB. If the application's business use case is small and we are not utilizing most of the packages that are part of System.Web assembly, we even have to use it to build and run the application. Secondly, the application can only be hosted on IIS.

With .NET Core, that System.Web assembly is split into modules and provides a smaller footprint as compared to the .NET framework. Secondly, there is no tight coupling with IIS. The application can run cross-platform on Kestrel, which is a lightweight server, whereas the HTTP request/response pipeline also can be defined using OWIN (Open Web Interface for .NET) middleware (see Figure 4-22).

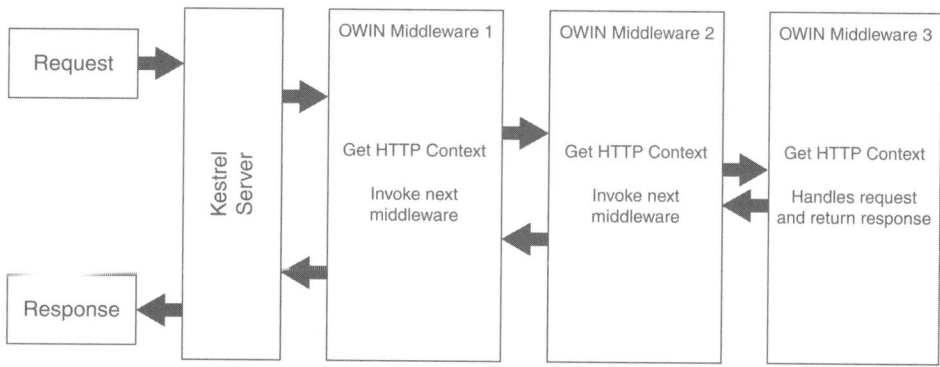

FIGURE 4-22   OWIN pipeline in ASP.NET Core

Figure 4-22 shows an OWIN pipeline in ASP.NET Core. The pipeline contains three middleware items defined in the Configure method of the Startup class as OWIN Middleware 1, OWIN Middleware 2, and OWIN Middleware 3. When the request comes to the Kestrel service, the first middleware (OWIN Middleware 1) is invoked. It can perform some operation and then call the next middleware in the pipeline, OWIN Middleware 2, as defined in sequence. OWIN

Middleware 2 performs some operation and calls the third middleware—OWIN Middleware 3. Finally, OWIN Middleware 3 handles the request and returns the response.

> **NEED MORE INFO?** **OWIN specification**
>
> See the OWIN specification at *http://owin.org*

We will use the ASP.NET Core Web API model to create a project for the payment service. You can use VS Code or Visual Studio IDE itself to create this project. The example here shows the Visual Studio steps to create a new ASP.NET Core project for the payment service API.

1. Open Visual Studio and choose the option to create a new project. Then search for **asp. net core** and select **ASP.NET Core Web Application,** as shown in Figure 4-23.

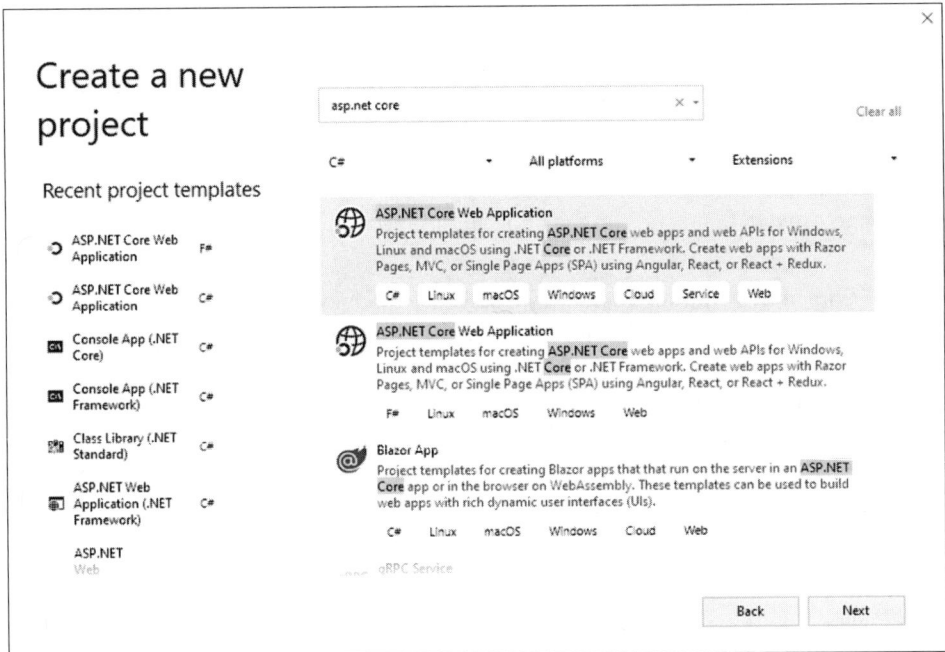

FIGURE 4-23 Choosing an ASP.NET Core Web Application project when creating a new project in Visual Studio 2019

2. Name the solution AuctionPayment and set the project name as PaymentService. Choose ASP.NET Core Version 3.1 and select an API template for the project. See Figure 4-24.

Once the project is created, build and run the project. By default, the API project contains one controller named Values. However, we will be building our own controller to expose methods to create and get payments.

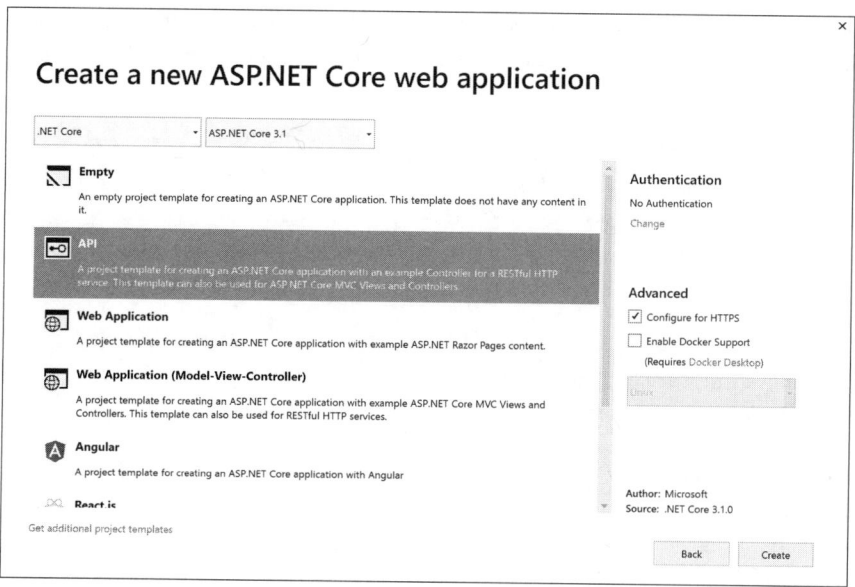

FIGURE 4-24   Selecting API as the application model for the ASP.NET Core Web Application

Create a code-first model using EF Core

In .NET, there are various options to perform CRUD operations. The *ADO.NET* library is one of the primitive libraries for performing database operations. However, ORM (Object Relational Mapping) frameworks enable developers to work with databases using .NET Objects. There are several ORM frameworks such as Ibatis.NET and NHibernate that can be used with .NET. However, the Entity Framework Core is a very lightweight, extensible, and open-source framework that provides various models as discussed below:

- **Database First Model**   If you already have a database in place, the code can be generated using it.
- **Code First Database**   Define POCO (Plain Old CLR Object) classes to model the data and use a Fluent API or declarative approach to design the mode using a code-first approach. Based on that, the database is generated.

For the Payment Service, we will use a code-first approach:

1. First, we need to add the EF Core assemblies into our project. Add following two libraries from *Nuget.org*:
   - `Microsoft.EntityFrameworkCore`
   - `Microsoft.EntityFrameworkCore.SqlServer`

2. After adding the above packages, we will create a new Models folder at the root of the project and create a class named `AuctionPayment`. Listing 4-12 shows the code snippet for the AuctionPayment class:

LISTING 4-12 Auction Payment entity

```
namespace PaymentService.Models
{
    public class AuctionPayment
    {
        [Key]
        public int Id { get; set; }
        public string CreditCardNo { get; set; }
        public string Name { get; set; }
        public int IdAuction { get; set; }
        public string BidUser { get; set; }
        public int Month { get; set; }
        public int Year { get; set; }
        public int PaymentStatus { get; set; }
        public DateTime PaymentDate { get; set; }

    }
}
```

3. The `AuctionPayment` class contains ID as the primary key that holds the payment trans-action ID. Other fields such as `CreditCardNo` hold the credit card number of the payer. The `Name` field holds the name of the payer, `idAuction` holds the Auction ID of the Auction table, and so on.

4. Let's now add a connection string key in the `AppSettings.json` file that points to the database created in Listing 4-12. The database connection string can be obtained from the Azure portal, as shown in Listing 4-13.

LISTING 4-13 Adding connection string for Auction Payment database

```
{
  "Logging": {
    "LogLevel": {
      "Default": "Warning"
    }
  },
  "AllowedHosts": "*",
  "ConnectionStrings": {
  "PaymentServiceContext": "Server=tcp:yourserver.database.windows.net,1433;Initial Catalog=
  AuctionPaymentDB;Persist Security    Info=False;User ID=sqladmin;Password=P@ssw0rd!@#;Multipl
  eActiveResultSets=False;Encrypt=True;TrustServerCertificate=False;Connection Timeout=30;"
  }
  }
```

5. Specify your server name, database name, user ID, and password for the Payment-ServiceContext key.

Create payment service context

Create a new class under the Data folder and name it `PaymentServiceContext`. This class should be derived from the `DbContext` class. This is the entry point where we specify the models needed to be considered for table creation. We can also specify and manipulate configuration values for the data context using Fluent API.

We can add the `AuctionPayment` model using `DbSet`. Any model for which we need to create a corresponding table should be added with `DbSet` type. Listing 4-14 shows the complete code snippet of the `PaymentServiceContext` class.

LISTING 4-14 Payment Service Context class

```
namespace PaymentService.Models
{
    public class PaymentServiceContext : DbContext
    {
        public PaymentServiceContext (DbContextOptions<PaymentServiceContext> options)
            : base(options)
        {
        }

        public DbSet<PaymentService.Models.AuctionPayment> AuctionPayment { get; set; }
    }
}
```

After writing the database context class, we can use EF Core CLI commands to create the database. To do this, open a command prompt and navigate to your project folder. We need to first run the following command to create a migration:

```
dotnet ef migrations add Initial
```

This command will search for the class that is derived from the `DbContext` and read the types that are specified as `DbSet` and create a class for migration. The database connection string is automatically picked up if the name of the `ConnectionString` is the same as the `DbContext` class name. In our case, the `DbContext` class is `PaymentServiceContext`, and we specify the same connection string key as `PaymentServiceContext`, so the same connection string will be used. Once the above command is executed, you will see the output shown in Figure 4-25.

FIGURE 4-25 Migrations output

Listing 4-15 shows the `Up` and `Down` methods of the Initial class that contains the object field mapping with the table column. The `Up` method is used to apply the migration, and the `Down` method is used to roll back the migration.

LISTING 4-15   Initial migration class

```
public partial class Initial : Migration
{
    protected override void Up(MigrationBuilder migrationBuilder)
    {
        migrationBuilder.CreateTable(
            name: "AuctionPayment",
            columns: table => new
            {
                Id = table.Column<int>(nullable: false)
                    .Annotation("SqlServer:ValueGenerationStrategy", SqlServerValue-
                    GenerationStrategy.IdentityColumn),
                CreditCardNo = table.Column<string>(nullable: true),
                Name = table.Column<string>(nullable: true),
                IdAuction = table.Column<int>(nullable: false),
                BidUser = table.Column<string>(nullable: true),
                Month = table.Column<int>(nullable: false),
                Year = table.Column<int>(nullable: false),
                PaymentStatus = table.Column<int>(nullable: false),
                PaymentDate = table.Column<DateTime>(nullable: false)
            },
            constraints: table =>
            {
                table.PrimaryKey("PK_AuctionPayment", x => x.Id);
            });
    }

    protected override void Down(MigrationBuilder migrationBuilder)
    {
        migrationBuilder.DropTable(
            name: "AuctionPayment");
    }
}
```

Next, we can apply the migration to the database by running the following command:

```
dotnet ef database update
```

With code-first model, the state of migration is maintained on the target database itself. When the migration is run for the first time, a new __EFMigrationsHistory table is created in the database that holds the information about the migrations that were applied. When the database update is called, it checks the last-applied migration and applies all the remaining migrations that do not exist in the __EFMigrationsHistory table. You can also specify the specific migration while running the database update command as follows:

```
dotnet ef database update -migration {name_of_migration}
```

The `name_of_migration` should only contain the migration name, not the complete alpha-numeric value. For example, for the `Initial` migration we ran previously, we created a file named 20191007144745_Initial. Here, the migration name will be just Initial.

To roll back the migration, we can run the following command and specify the name of the migration.

```
dotnet ef database update Initial
```

Create an auction payment controller

So far, we have created a .NET Core project and created a database and table to hold auction payment transactions. Now we will add a new API Controller to expose certain methods to create, read, and update payment transactions. Table 4-5 shows the methods we will create in this section.

TABLE 4-5 Auction Payment API methods

API Method	HTTP Verb	Signature	Description
Get Auction Payments	GET	/api/auctionPayments	Returns the list of auction payments
Get Auction Payment by Id	POST	/api/auctionPayments/Id	Returns the auction payment based on the payment ID
Create Auction Payment	POST	/api/auctionPayments	Creates the auction payment

Create a new controller and name it `AuctionPaymentsController`, as shown in Listing 4-16.

LISTING 4-16 Auction Payments controller

```
namespace PaymentService.Controllers
{
    [Route("api/[controller]")]
    [ApiController]
    public class AuctionPaymentsController : ControllerBase
    {
        private readonly PaymentServiceContext _context;

        public AuctionPaymentsController(PaymentServiceContext context)
        {
            _context = context;
        }

        // GET: api/AuctionPayments
        [HttpGet]
        public async Task<ActionResult<IEnumerable<AuctionPayment>>> GetAuctionPayment()
        {
            Logger.Instance.LogMessage("Getting auction payment");
            return await _context.AuctionPayment.ToListAsync();
        }
}
```

```csharp
// GET: api/AuctionPayments/5
[HttpGet("{id}")]
public async Task<ActionResult<AuctionPayment>> GetAuctionPayment(int id)
{
    Logger.Instance.LogMessage($"Getting auction payment by ID where ID is {id}");
    var auctionPayment = await _context.AuctionPayment.FindAsync(id);
    if (auctionPayment == null)
    {
        return NotFound();
    }

    return auctionPayment;
}

// POST: api/AuctionPayments
[HttpPost]
public async Task<ActionResult<AuctionPayment>> PostAuctionPayment(AuctionPayment
auctionPayment)
{
    Logger.Instance.LogMessage($"Insert auction payment where Auction ID is
    {auctionPayment.IdAuction}");
    _context.AuctionPayment.Add(auctionPayment);
    await _context.SaveChangesAsync();

    return CreatedAtAction("GetAuctionPayment", new { id = auctionPayment.Id },
    auctionPayment);
}

private bool AuctionPaymentExists(int id)
{
    return _context.AuctionPayment.Any(e => e.Id == id);
}
    }
}
```

In Listing 4-15, we first injected the PaymentServiceContext object to perform database operations. There are two methods: HTTP GET return the auction payment(s), and the HTTP POST method creates an auction payment.

Developing an application front-end

Today, progressive web applications are very demanding when building a web-based frontend for any application. A progressive web application is built around common web technologies, such as JavaScript , HTML, and CSS, and runs on web or mobile platforms.

There are various client-side frameworks, such as React and Angular, that support progressive web development. We will develop the OAS frontend using Angular, which is a Typescript-based, open-source web application framework that is driven by the Angular Team at Google that provides a component-based model, which helps you to divide your SPA into various components and then associate HTML views or angular term templates while creating them. Some of the characteristics of Angular are as follows:

- In addition to being able to develop web applications, you also can create native mobile applications and desktop applications that can run across Windows, Linux, and Mac operating system with the support of the respective OS (Operating System) APIs.

- You can use TypeScript to write code. TypeScript is a superset of JavaScript developed by Microsoft that extends JavaScript by adding types to the language. Angular allows you to write code in TypeScript and converts that into JavaScript when it is compiled.

- Angular also turns the templates into highly optimized code.

- Provide a simple syntax for creating views or pages.

- Provides a CLI (Command Line Interface) to create a project or add components, services, and so on by running respective commands.

Prerequisites

The following tools must be installed to develop the front-end application.

- Visual Studio Code
- Node.js

Creating a front-end application

Once we have the Node.JS installed, we can install the Angular CLI by running the following command:

```
npm install -g @angular/cli
```

After running the above command, we can now create applications in Angular. Create a new folder where you want to create the Angular project and run the following command to scaffold the basic Angular project:

```
ng new onlineauctionweb
```

While running above command from the command prompt, it will ask you to add the Angular routing module, select **Yes**, and for **Stylesheets**, select **CSS** and press Enter. A basic Angular project will be scaffolded that can be run using the forllowing command, which will start the application and listen on the default port of 4200:

```
ng serve
```

Figure 4-26 shows the angular application running and listening for requests on port 4200.

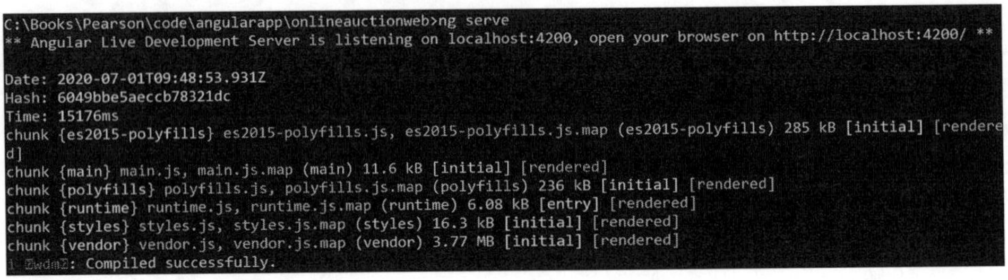

```
C:\Books\Pearson\code\angularapp\onlineauctionweb>ng serve
** Angular Live Development Server is listening on localhost:4200, open your browser on http://localhost:4200/ **

Date: 2020-07-01T09:48:53.931Z
Hash: 6049bbe5aeccb78321dc
Time: 15176ms
chunk {es2015-polyfills} es2015-polyfills.js, es2015-polyfills.js.map (es2015-polyfills) 285 kB [initial] [rendere
d]
chunk {main} main.js, main.js.map (main) 11.6 kB [initial] [rendered]
chunk {polyfills} polyfills.js, polyfills.js.map (polyfills) 236 kB [initial] [rendered]
chunk {runtime} runtime.js, runtime.js.map (runtime) 6.08 kB [entry] [rendered]
chunk {styles} styles.js, styles.js.map (styles) 16.3 kB [initial] [rendered]
chunk {vendor} vendor.js, vendor.js.map (vendor) 3.77 MB [initial] [rendered]
ℹ ｢wdm｣: Compiled successfully.
```

FIGURE 4-26 Running the Angular app

NOTE   **Port switch**

You can access the application from port *4200*. If the port is used by any other application, you can run the ng serve command with the --port switch as shown below:

ng server --port 80

Understanding the Angular project structure

In the Angular project, note the angular.json file, which contains all the configuration settings
of the project. It is placed at the root level of Angular project and is used when you build or run
your application.

If you open the angular.json file, you will notice that there is a main attribute defined in the
architect section. This main attribute specifies the file that bootstraps the angular application,
as shown in the Listing 4-17:

LISTING 4-17   Angular.json file

```
"architect": {
        "build": {
          "builder": "@angular-devkit/build-angular:browser",
          "options": {
            "outputPath": "dist/onlineauction-app",
            "index": "src/index.html",
            "main": "src/main.ts",
            "polyfills": "src/polyfills.ts",
            "tsConfig": "src/tsconfig.app.json",
            "assets": [
              "src/favicon.ico",
              "src/assets"
            ],
            "styles": [
              "src/styles.css",
              "node_modules/bootstrap/dist/css/bootstrap.min.css"
            ],
```

```
      "scripts": [],
      "es5BrowserSupport": true
    },
```

Let's open the `main.ts` file residing in the root of the angular project (see Listing 4-18).

LISTING 4-18   Main typescript

```typescript
import { enableProdMode } from '@angular/core';
import { platformBrowserDynamic } from '@angular/platform-browser-dynamic';

import { AppModule } from './app/app.module';
import { environment } from './app/environments/environment';

if (environment.production) {
  enableProdMode();
}


platformBrowserDynamic().bootstrapModule(AppModule)

  .catch(err => console.error(err));
```

This file imports some libraries and core angular packages and then specifies the main bootstrap module as `AppModule`. The `AppModule` is the main module of the application. It contains all the modules or components we need to use globally. For example, if we want to use the `HttpClientModule` to call services, we need to add this under the `@NgModule` section of this file.

Listing 4-19 shows the complete `app.module`. Several other modules have been added in the import section that are used in the OAS application.

LISTING 4-19   App module typescript

```typescript
import { BrowserModule } from '@angular/platform-browser';
import { NgModule } from '@angular/core';
import { FormsModule }   from '@angular/forms';
import { AppRoutingModule } from './app-routing.module';
import { AppComponent } from './app.component';
import { HttpClientModule } from '@angular/common/http';
import { AngularFontAwesomeModule } from 'angular-font-awesome';
import { StorageServiceModule} from 'angular-webstorage-service';
import { MsalService }   from './services/msal.service';
import {SecurityGaurdService} from '../app/services/securitygaurd.service'

@NgModule({
  declarations: [
    AppComponent
  ],
  imports: [
    BrowserModule,
```

```
      FormsModule,
      AppRoutingModule,
      HttpClientModule,
      AngularFontAwesomeModule,
      StorageServiceModule,

    ],
    providers: [MsalService, SecurityGaurdService] ,
    bootstrap: [AppComponent]
})
export class AppModule { }
```

The boostrap tag holds the value of the main Angular component that binds the template as the landing page when the application is accessed by the user. You can open the app.component and see the respective HTML page associated with this component. The templateUrl specifies the actual path to the HTML view associated Listing 4-20 shows complete code snippet of the AppComponent typescript file.

LISTING 4-20   App component typescript

```
import { Component, Inject, OnDestroy, OnInit } from '@angular/core';
import { componentNeedsResolution } from '@angular/core/src/metadata/resource_loading';
@Component({
  selector: 'app-root',
  templateUrl: './app.component.html',
  styleUrls: ['./app.component.css']
})
export class AppComponent   {
  title = 'Online Auction System';
}
```

The selector is specified as app-root. In Angular, we can use this selector in any of the HTML pages to render the respective view page as specified in the templateUrl attribute. Let's define the selector tag in the Index.html file residing at the root of the src folder. Following is the body tag of Index.html file that contains the app-root tag, as shown below:

```
<body>
  <app-root></app-root>
</body>
```

Figure 4-27 shows the overall project structure that contains an app folder. Inside the app folder, we have angular components inside the components folder; environment-specific files that hold key/value pairs are in the environments folder; and services that are used by the components are inside the services folder.

The next section contains more about the core concepts of Angular.

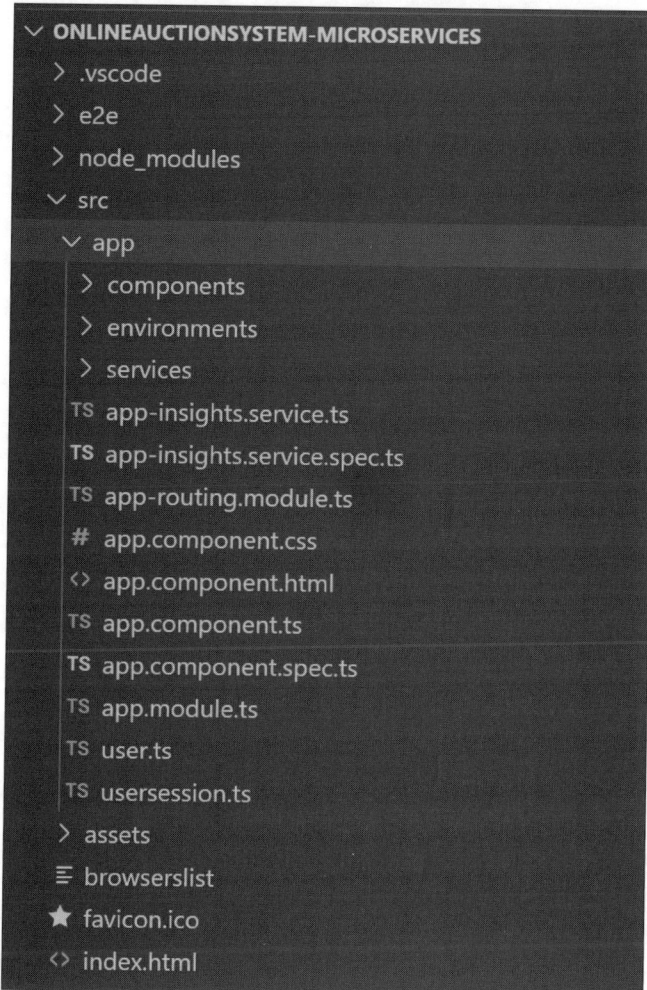

FIGURE 4-27 The project structure of the OAS Angular app

Angular concepts

Before starting to develop the OAS front-end application on Angular, you need to understand the key building blocks on which the Angular framework relies (see Figure 4-28).

Following are some of the core fundamentals you need to understand when building application on Angular.

- Module
- Component
- Template
- Service
- Routings

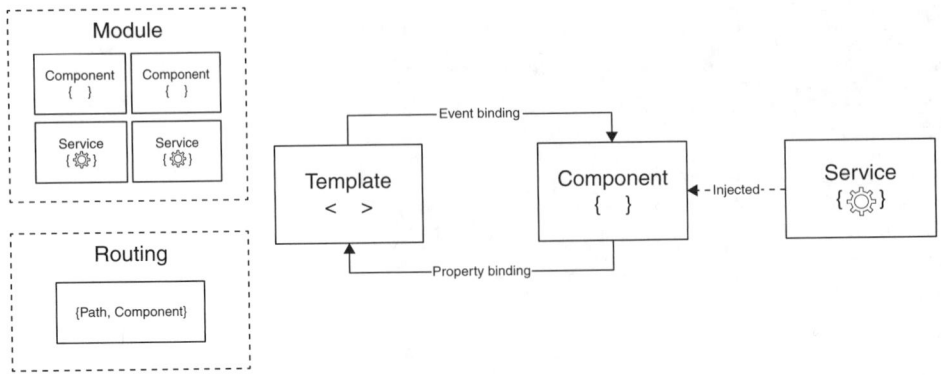

FIGURE 4-28   Key building blocks of Angular

Module

A module, also known as @NgModule in Angular, is a collection of components, services, directives, and so on. By default, the Angular application has one root module named AppModule, but we can separate the application into various modules and load them as they are needed. This concept is also known as "lazy loading." To learn more about lazy loading modules in Angular, see *https://angular.io/guide/lazy-loading-ngmodules*.

If the application is large, it is always a better practice to split it into various modules and load each module and related components and services when they are needed. Suppose you are developing an ERP (Enterprise Resource Planning) application that contains modules such as human resource management, payroll system, finance system, and others. Each of these modules can be segregated into various *angular modules*, and depending on the user's access, related components, services, directives, and other artifacts can be loaded.

Component

The component contains the actual application logic and defines a class that contains a selector to use that component, as well as the association with the respective HTML template using the templateUrl attribute that displays the HTML view where the selector is used. The @Component directive is used to declare a class as a component, and the selector tells Angular to render the HTML page as specified in the templateUrl. For example, in the code snippet below, the selector is alert, so if a page contains a selector like the <alert> tag, it will replace that tag with the associated HTML template, as specified in the templateUrl, which is alert.component.html.

```
import {Component, OnInit } from '@angular/core';
@Component({
  selector: 'alert',
  templateUrl: './alert.component.html'
})
export class AlertComponent implements OnInit{
}
```

Template

A template is an HTML file that is associated to the component and renders on a page where the component selector is used. It provides two types of binding: property binding and event binding.

- **Property binding**   The property binding is used to update the value in the HTML where the actual template is rendered (through a selector). For example, to update template data, we can use property binding. The component member variables can be used on the template itself and can be updated from the associated component.

- **Event binding**   The event binding is used to invoke the component methods to implement some back-end logic. For example, suppose a template contains a form and a button to save information. When the user submits the form by clicking the **Save** button, the respective method will be invoked in the component. This can be achieved using an event binding.

Service

Some classes in Angular need to be used across the whole application. For example, the Security service enables the security flows in the application and provides options to register or log in to the system. The HTTP service is another example that can be used to perform HTTP operations.

A service class can be injected into the component constructor using DI (dependency injection). However, this service class should be decorated with the `@Injectable()` decorator.

Routing

The Routing module in Angular is used to define the routing paths. You can define the routes in an Angular application in the `app-routing.module`. It provides an array of routes that define the collection of routes that contain a path and a component as one of the key variables. Here is the sample routing defined for the OAS application:

```
const routes: Routes = [
  { path: 'home', component: HomeComponent }
];
```

Developing a security module

First, the user needs to register in order to use the OAS. We are leveraging the Azure AD B2C to enable the self-registration process in the system. Once the user is registered and logged in, we save the user ID in the session storage object and then use it to display the user information on the UI. Also, we use it when performing database operations to record which user has taken a particular operation.

In this section, we will enable user authentication using Azure AD B2C. Azure AD B2C is an identity management service that allows users to self-register and log in to the system. In Azure AD B2C, you have to register an application on an Azure AD B2C tenant and choose the predefined user flows, which includes sign in, sign up, profile editing, password reset, and others. For each user flow, you can select the user attributes, which become form fields for that user flow page. The user will be required to enter that information when registering from the registration page.

Custom fields can also be added and selected as part of the user flow. The benefit of Azure AD B2C is that it provides a configuration-based model, and as a developer, you just need to configure user attributes, and the UI is automatically generated when you access that user flow based on the fields you have chosen while creating that user flow. Moreover, you can also define claim attributes, which are also user fields that are entered by the user while registering. The claim attributes selected in the configuration will be part of the token generated by the Azure AD B2C tenant. To learn more about Azure AD B2C and configure user flows, refer to the Chapter 7, "Security in microservices."

The following command creates a new service file named `msal.service` in the services folder. To implement an authentication scenario with Azure AD B2C, Microsoft provides the MSAL (Microsoft Authentication Library) SDK for various platforms.

```
ng generate service msal
```

For angular, we can add this library by running the following command:

```
npm install msal @azure/msal-angular –save
```

After adding this library, import the classes in the `msal.service`, as shown below:

```
import { Injectable } from '@angular/core';
import * as Msal from 'msal';
```

The `Injectable` is used in the service class `MsalService`, as shown in the code snippet below so that this `MsalService` can be used and injected in other Angular components.

LISTING 4-21   MSAL service class

```
@Injectable()
export class MsalService {

    B2CTodoAccessTokenKey = "b2c.access.token";
    tenantConfig = {
        tenant: "oasmicroservics.onmicrosoft.com",
        // Replace this with your client id
        clientID: '3ce6224e-2812-44fb-93f5-f24e84ebad6a',
        signInPolicy: "B2C_1_signupsingin1",
        signUpPolicy: "B2C_1_signupsingin1",
      //redirectUri:"https://onlineauctionapp.azurewebsites.net/main",
        redirectUri:"http://localhost:4200",
        b2cScopes:["https://oasmicroservics.onmicrosoft.com/oas/user_impersonation","https://
        oasmicroservics.onmicrosoft.com/oas/read openid offline_access"]
    };


    // Configure the authority for Azure AD B2C
```

```
authority = "https://oasmicroservics.b2clogin.com/tfp/" + this.tenantConfig.
tenant + "/" + this.tenantConfig.signInPolicy;
```

In Listing 4-21, we have defined the `tenantConfig` object that contains some key/value pairs, such as the tenant address, which is the Azure AD B2C tenant address; the `ClientID` that you can obtain after registering an application on Azure AD B2C; the `signInPolicy` and `signUp-Policy` names that should refer to the user flows you created in Azure AD B2C; and the `redirectUri` that points to the actual URL of your front-end application and the `b2cScopes`. Lastly, the authority should be a complete URI to connect to the authority server.

After specifying the `tenantConfig` object, we initialize the client application object to pass these values:

```
clientApplication = new Msal.UserAgentApplication(
    this.tenantConfig.clientID, this.authority, this.callbackHandler,
    {validateAuthority:false, redirectUri: this.tenantConfig.
    redirectUri, cacheLocation:"sessionStorage" }
);
```

In the above code, we initialized the `clientApplication` by initializing the `UserAgentApplication` of MSAL library and passing the `ClientID`, authority URL, and other details. The `callbackHandler` is a delegate to a function in the same MSAL component class that will be called once the successful authentication is done. Following is the implementation of the `callbackHandler`:

```
public callbackHandler (errorDesc: any, token: any, error: any, tokenType: any) {
    sessionStorage.clear();
    sessionStorage.setItem(this.B2CTodoAccessTokenKey, token);
    console.log("token has been saved");
    sessionStorage.setItem("Token",token);
}
```

Next, we add the log in, sign up, and log out methods in the MSAL component class so we can call them from the `app.component` class, choose where the associate HTML view for the `app.component` will expose links to register, and log in to or log out of the OAS. Listing 4-22 shows the methods we will define in the MSAL component class.

LISTING 4-22   MSAL service methods

```
// To register to the OAS
    public signup():void{
        this.clientApplication.authority = "https://oasmicroservics.b2clogin.com/tfp/" + this.
        tenantConfig.tenant + "/" + this.tenantConfig.signUpPolicy;
        this.authenticate();
    }
// To login to the OAS
public login():void{
        this.clientApplication.authority = "https://oasmicroservics.b2clogin.com/tfp/" + this.
        tenantConfig.tenant + "/" + this.tenantConfig.signInPolicy;
```

```
        this.authenticate();
    }
  //To log out from the OAS
    logout(): void {
        this.clientApplication.logout();
    };
```

There are few other methods as well to return the user name, user email and user Id in the session storage.

```
  //Returns User Object
    getUser(){
        return this.clientApplication.getUser();
    }
  //Returns User Name
    getUserName():string {
        return this.getUser().name;
    }
  //Returns User Email
    getUserEmail(): string{
        return this.getUser().idToken['emails'][0];
    }
  //Returns User Id
    getUserId(){
        return this.getUser().userIdentifier;
    }
```

Now modify the app.component to expose some methods to be used when the user clicks **Login** or **Register** from the home page. Also, we need to inject the MSAL service in the constructor so we can call the respective log in, log out, and sign-up methods. Here is the updated constructor of app component that injects the session storage and application insights object:

```
  constructor(private msalService: MsalService,
    @Inject(SESSION_STORAGE) private storage: WebStorageService, private
    appInsights: AppInsightsService)
{ }
```

> **NOTE**   **Application Insights**
>
> Application Insights is used throughout the OAS to write custom telemetry information.

Listing 4-23 shows the rest of the methods used to read the user's email address, username, user identifier (user ID) and contains methods to sign up, log in and log out.

LISTING 4-23   App component methods

```
    userEmail() {   return this.useremail; }
    userName() {
if(this.msalService.isLoggedIn()){
                    this.appInsights.instance.setAuthenticatedUserContext(this.msal-
                    Service.getUserEmail(), "User");
```

```
            }
    else{
                        this.appInsights.instance.clearAuthenticatedUserContext();
        }
        return this.msalService.getUserName();
    }
    userIdentifier(){ return this.storage.get("userId"); }
    login(){ this.msalService.login(); }
    signup(){ this.msalService.signup();   }
    logout(){ this.msalService.logout();   }
```

We added another method named as IsUserLoggedIn to check whether the user is logged in and if so, show the **Logout** option. Otherwise, if the user is not logged in, the **Register** and **Login** buttons will be displayed on the home page.

```
isUserLoggedIn(){
    if(this.msalService.isLoggedIn()){
        this.useremail= this.msalService.getUserEmail();
        this.storage.set("userId", this.msalService.getUserId());
        this.storage.set("userName", this.msalService.getUserName());
        this.storage.set("userEmail", this.msalService.getUserEmail());
        return true;
    } else return false;
    }
}
```

Finally, in the app component view, we have used these HTML div elements to show **Login**, **Register**, and **Logout**, options, and used ngIf (Angular If) directive to invoke the IsUser-LoggedIn method and show or hide the div based on the result.

```
            <div class="center" *ngIf="!isUserLoggedIn()">
                <button (click)="login()" class="btn btn-default btn-sm"
                style="float:right;">Login</button>
                <button (click)="signup()" class="btn btn-default btn-sm"
                style="float:right;">Register</button>

            </div>
            <div class="center" *ngIf="isUserLoggedIn()">
                <button (click)="logout()" class="btn btn-default btn-sm"
                style="float:right;">Logout</button>
                <button  class="btn btn-default btn-sm" style="float:right;">Welcome
                {{userName()}}</button>
            </div>
```

So far, the basic authentication framework is ready, once the user clicks **Login** or **Register** from the home page, he or she will be redirected to the authority server page, which provides a challenge to enter a username and password. On a successful authentication, the page will be redirected to the OAS site URL with a valid token. The callbackHandler method will be invoked, and a valid token will be retrieved and saved in the session storage. Also, on the homepage that is associated with the app component page, the `IsUserLoggedIn` method will be invoked, which calls the `isLoggedIn` method to determine whether the user is successfully logged in. Next, it pulls information specific to the user ID, username, and user email and saves it to the session storage object.

If the user directly accesses the page through the routing path, that component page can be loaded, and the respective operation can be performed. Route Guard in Angular can be used to protect pages and provide authorized access depending on whether the user is logged in.

Route Guard tells the router whether the navigation is allowed by reading a *Boolean* value. If the Boolean value is `true`, the path can be navigated; if the Boolen value is `false`, it cannot be navigated. There are five types of Route Guards available in Angular: `CanActivate`, `CanActivateChild`, `CanDeactivate`, `CanLoad`, and `Resolve`. We will use `CanActivate` to enable protected access to the other pages on the OAS.

> **NEED MORE REVIEW?**   **Route Guards**
>
> To learn more about them, see *https://angular.io*.

To work with Route Guard, we first add a new service and name it `SecurityGaurdService`. Use the command for creating service in Angular as shown below.

```
ng generate service SecurityGaurd
```

Next, import the following namespaces in the `SecurityGaurdService`:

```
import { Injectable } from '@angular/core';
import { CanActivate, ActivatedRouteSnapshot, Router } from '@angular/router';
import { MsalService }  from './msal.service';
```

Implement the `CanActivate` interface on the `SecurityGuardService` and provide the implementation of the `CanActivate` method as shown below. We also need to inject the `MsalService`, so use a parameterized constructor:

```
@Injectable()
export class SecurityGaurdService implements CanActivate {
  constructor(private _router:Router, private msalService: MsalService) { }
  canActivate(route:ActivatedRouteSnapshot): Boolean
{
    if(this.msalService.isLoggedIn())
  {
 return true;
  }
   else {  return false; }
  }
}
```

The above `CanActivate` method is calling the `isLoggedIn` method of `MsalService` that returns the Boolean value and returns `true` if the user is logged in; otherwise, it returns `false`.

To enable Security Guard for the particular routing path, we can modify the `app.routing.module` and specify the `canActivate` property followed with the name of the Security Guard service you just created.

```
const routes: Routes = [
    { path: 'home', component: HomeComponent, canActivate:[SecurityGaurdService]  },
    { path: 'auction', component: AuctionComponent, canActivate:[SecurityGaurdService] },
    { path: 'bid', component: BidComponent, canActivate:[SecurityGaurdService] },
    { path: 'mytransactions', component: MyTransactionsComponent, canActivate : [Security
       GaurdService] },
    { path: 'alert', component: AlertComponent},
    { path: 'main', component: MainComponent},
    { path: 'biddetail', component: BidDetailComponent, canActivate : [SecurityGaurd
       Service] },
    { path: 'winningbids', component: WinningBidsComponent, canActivate: [SecurityGaurd
       Service]},
    { path: 'makePayment', component: PaymentComponent, canActivate: [SecurityGaurd
       Service]},
    { path:'**'  , redirectTo:'main'}
];
```

Until a user is successfully logged in, the routing path that has the `SecurityGuardService` defined for the `canActivate` property will not be accessible.

Configuring environment files

In any application, it is always a better practice to keep the constant values in a common class or file that can be read throughout the application. In a front-end application, we will be calling various microservices to perform specific operations for a particular subdomain. Hence, using their URLs directly in the component classes is not easily manageable.

By default, when the angular project is created, two files are added into the environments folder: `environment.ts` and `environment.production.ts`. The `environment.ts` file is used for keeping the development-related configuration, whereas the `environment.production.ts` file is used to keep the production-related configuration. Let's add the constants in both the environment files as follows. Specify the values as per your configuration.

Here is the environment file for development:

```
export const environment = {
  production: false,
"auctionAPI":"https://oasdev.azure-api.net/auction",
"bidAPI":"https://oasdev.azure-api.net/bid",
"paymentAPI": "https://oasdev.azure-api.net/payment"
};
```

Here is the environment file for production:

```
export const environment = {
  production: true,
```

```
"auctionAPI":"https://oasprod.azure-api.net/auction",
"bidAPI":"https://oasprod.azure-api.net/bid"
"paymentAPI": "https://oasprod.azure-api.net/payment"
};
```

Develop the create auction form

The create auction form is the first component we will be creating in this section to let users create auctions. The create auction form contains fields such as Auction Name, Description, Staring Bid Value, Active For Hours, and Auction Image. With angular CLI, components can easily be created using the following command:

```
ng generate component auction
```

This command will create a folder inside a component named `auction` and will create component and html files. In the `auction.component.html` file, we need to create a form that contains respective fields and a button to submit the form fields to the respective action method, as defined in the associated `auction.component`. To work with forms in Angular, we can use NgForm.

The NgForm is a directive exported from `FormsModule`, which at runtime, automatically creates an HTML `<form>` element. To use NgForm, we need to add into the `imports` array of the `@NgModule` in app.module.

Listing 4-24 shows the complete HTML code of the `auction.component` HTML file:

LISTING 4-24   Create Auction HTML

```
<div class="container">
<h3>
  <img width="30" alt="Create Auction" src="assets/images/createauction.
  png" />  Create Auction</h3>
    <form #f="ngForm" (ngSubmit)="createAuction()">
      <p>
      <alert></alert>
      </p>
      <hr>

    <div class="form-group">
    <label for="name">Auction Name</label>
        <input type="text"   [(ngModel)]="auction.name"
        name="name"  class="form-control" id="name" required
        #name="ngModel">

        <div *ngIf="name.invalid  && (name.dirty || name.touched)"
            class="alert alert-danger">
          Name is required
        </div>
      </div>
```

```html
    <div class="form-group">
      <label for="description">Describe something about this auction...</label>
      <textarea [(ngModel)]="auction.description" name="description"  class="form-con-
      trol" rows = "3" cols = "60" id = "description" required #description="ngModel"></
      textarea>
      <div  *ngIf="description.invalid  && (description.dirty || description.touched)"
          class="alert alert-danger">
           Description is required
      </div>
    </div>


    <div class="form-group">
      <label for="startingPrice">Starting Bid Value (in USD)</label>
     <input [(ngModel)]="auction.startingPrice"  type="text" class="form-control"
    name="startingPrice" id="startingPrice" required #startingPrice="ngModel">
      <div *ngIf="startingPrice.invalid && (startingPrice.dirty || startingPrice.
      touched)"
      class="alert alert-danger">
      Starting Price is required
  </div>
  </div>

  <div class="form-group">
    <label for="activeInHours">Active for (in Hours)</label>
    <input [(ngModel)]="auction.activeInHours"  type="text" class="form-control"
    name="activeInHours" id="activeInHours" required #activeInHours="ngModel">
    <div  *ngIf="activeInHours.invalid && (activeInHours.dirty || activeInHours.touched)"
    class="alert alert-danger">
    Active for (in Hours) is required
</div>
</div>

  <div class="form-group">
     <image-component (fileUploadedEvent)="receiveImageEventHandler($event)" >
     </image-component>
   </div>

    <button  *ngIf="!loadingAuction" type="submit" class="btn btn-primary"
    [disabled]="!f.valid">Save </button>

    <div class="d-flex justify-content-center">
      <div class="spinner-border text-secondary"
      style="width: 3rem; height: 3rem;" role="status" *ngIf="loadingAuction">
        <span class="sr-only">Loading...</span>
      </div>
```

```
    </div>
  </form>
</div>
```

In the above HTML, we have defined a template reference variable called as #f that references the ngForm directive of the create auction form. In the ngSubmit directive, we specified the createAuction method, which is defined in the auction component class. Finally, we bound each field with the ngModel to get the data when the createAuction method is invoked, and the field value was read using the auction object created in the initialization of the auction component. Alternatively, we can also pass the form data directly when calling the createAuction method and passing the template reference variable f as a parameter.

Listing 4-25 shows the createAuction method defined in the auction component class that is reading the form fields values using the local auction object bind through the ngModel directive, setting the authorization token (because the services are protected, and we need to pass a valid token while making POST request) and then make an HTTP Post request to the auction service using the httpClient object.

LISTING 4-25   Create Auction method

```
public createAuction() {

  console.log("Token value from Auction Component " +this.msalService.getToken());
   // console.log("token is "+ this.storage.get("token"));
  this.loadingAuction=true;
  console.log("file is"+ this.auction.image);
  console.log("User Id is "+ this.storage.get("userId"));
  this.auction.userId=this.storage.get("userId");
  this.auction.userName = this.storage.get("userName");
  const headerSettings: {[name: string]: string | string[]; } = {};

  headerSettings['Authorization'] = 'Bearer ' + this.msalService.getToken();
  headerSettings['Content-Type'] = 'application/json';
  const newHeader = new HttpHeaders(headerSettings);

  this.httpClient.post(environment.auctionAPI+"/auctions", this.auction, {headers:
  newHeader}).subscribe(
          (res) => {

              var id = res;
              console.log(id);
              this.loadingAuction=false;
              this.alertService.add({"type":"success", "message": "Auction created
              successfully"});
              this.resetFields();

          },
```

```
    (err) => {
        this.loadingAuction=false;
        console.log(err);
        this.alertService.add({"type":"danger",
        "message": "Some error occured, please contact Administrator"});
        this.appInsights.instance.trackException(err);
    });

    console.log("Auction is created");

    this.appInsights.instance.trackEvent({name: 'CreatedAuction'});
    this.submitted = true;
}
```

In the code shown in Listing 4-25, we are reading the logged-in user ID from a session storage and setting the request headers, such as the `Authorization` header, to pass a bearer token and setting the `Content-Type` to send the payload in JSON format.

Next, we used the post method of `httpClient` object to make an `HTTP POST` request and pass the complete URI to call the post service. The auction service is hosted inside a container and configured behind an API gateway. The `environment.auctionAPI` is the key defined in the `environment.json` file to avoid hardcoding the DNS. The post method takes a URL and object and optionally, headers—if there are any—and subscribes to the callback event that receives the response or error from the service. On a successful operation, a success message is displayed. If an error occurs, an error message will be displayed, and an exception will be logged to Application Insights.

Developing an active auctions page

In this section, we will add a component to display active auctions. Use the same command discussed earlier to create a new component name `bid`.

We need to create a method of `getActiveBids` that can be used to fetch all the active auctions by making a call to the auction service. To do this, follow these steps:

1. First, add the parametrized constructor to inject objects such as `HttpClient`, `Router`, and so on.

```
    constructor(private httpClient: HttpClient, private alertService: AlertService,
private router: Router,
        private appInsights: AppInsightsService){}
```

2. Create a `getActiveBids` method that pulls information from the Auction database and sets it to the local auctions list. See Listing 4-26.

LISTING 4-26 Get Active Bids method

```
public getActiveBids() {
        this.loading = true;
        this.httpClient.get(environment.auctionAPI + "/auctions").subscribe(
            (res) => {
```

```
            this.auctions =res;
            console.log("Active auctions are loaded");
            console.log(res);
            this.loading = false;
        },
        (err) => {
            console.log(err);
            this.alertService.add({"type":"danger",
            "message":  "Some error occured, please contact Administrator" });
            this.loading = false;
            this.appInsights.instance.trackException(err);
        });
        this.appInsights.instance.trackEvent({name:'LoadedActiveBids'});
    }
```

Now specify the ngOnInit method, and call getActiveBids as shown below:

```
ngOnInit() {
    this.getActiveBids();
}
```

3. Next, we need to update the view page to show a table that displays the list of rows and columns for active auctions. Use the code shown in Listing 4-27 to create a table to display active auctions:

LISTING 4-27   Active Auctions HTML

```html
<div class="container">
  <table class="table table-hover " *ngIf="!loading">
    <thead>
      <tr>
        <th scope="col">#</th>
        <th scope="col">Name </th>
        <th scope="col">Description</th>
        <th scope="col">Starting Price</th>
        <th scope="col">Last Bid Price</th>
        <th scope="col">Remaining Time</th>
        <th scope="col">Created On</th>
        <th scope="col">Created By</th>
      </tr>
    </thead>
    <tbody>
      <tr *ngFor="let auction of auctions;" (click)="rowSelected(auction);">
        <th scope="row">{{auction.idAuction}}</th>
        <td>{{auction.Name}}</td>
        <td>{{auction.Description}}</td>
        <td>{{auction.StartingPrice | currency }}</td>
        <td>{{auction.LastBidOffer | currency }}</td>
```

```
            <td>{{auction.ActiveInHours}} hrs.</td>
            <td>{{auction.AuctionDate | date: 'dd/MM/yyyy h:mm a' }}</td>
            <td>{{auction.userName}}</td>
        </tr>
    </tbody>
  </table>
</div>
```

Developing a submit bid form

The bid detail form is used to bid against active auction. Create a new form, name it `biddetail`, and place it in the `biddetail` folder. Add a `biddetail` model class that contains properties used to serialize/deserialize the object:

```
export class BidDetail {
    constructor(
        public bidId: number,
        public auctionId: number,
        public bidAmount: number,
        public userId: string,
        public bidDate: Date,
        public userName: string
    ) {  }
}
```

The `biddetail.component` contains these methods:

- `loadBidByAuctionId`   Loads the auction detail to display on a bid page
- `loadPreviousBids`   Loads the last bids made on that auction
- `submitBid`   Submits a bid on that auction

Listing 4-28 shows the logic implemented for `loadBidByAuction`.

LISTING 4-28   Load Bid by Auction Id method

```
public loadBidByAuctionId(){
        this.httpClient.get(environment.auctionAPI+ "/auctionById/"+this.auctionId).
        subscribe(
            (res) => {
                            this.auction =res[0];
                var currentDateTime = new Date();
                var dt2 = new Date(this.auction.auctionDate);
                            var expiryDate =dt2;
                expiryDate.setHours(expiryDate.getHours() + (this.auction.activeInHours));
                var diffMs= expiryDate.getTime()- currentDateTime.getTime();
                diffMs= diffMs/60000;
                if(diffMs<0){
                this.loadMakeOffer=true;
                this.alertService.add({"type":"warning", "message": "Auction closed now!"});
```

```
    }else{
        this.elapsedTime=diffMs;
        this.loadMakeOffer=false;
        console.log("Total elapsed time left is"+ this.elapsedTime);
    }
    console.log("Total minutes left is "+ diffMs)
    this.imageFile  = this._sanitizer.bypassSecurityTrustResourceUrl
    ('data:image/jpg;base64,'
    + res[0].image);

},
(err) => {
    console.log(err);
    this.alertService.add({"type":"danger", "message":  "Some error occured,
    please contact Administrator"});
    this.appInsights.instance.trackException(err);
});
this.appInsights.instance.trackEvent({name:"LoadedBidbyAuctionID"});
}
```

In Listing 4-28, we called the `AuctionService` to get auction information by auction ID. Then, we bound the response to the local `auction` variable connected to the page to display auction information. To display the remaining time, we have implemented some logic to show the time left to submit a bid on that auction. To load previous bids. Listing 4-29 shows the `loadPreviousBids` method.

LISTING 4-29   `loadPreviousBids` method

```
public loadPreviousBids(){

    this.loadingPrevBids=true;
    this.httpClient.get(environment.bidAPI+ "/bid?auctionID="+this.auctionId).
    subscribe(
        (res) => {
            console.log("Response is "+ res);
            this.bids =res;
            this.loadingPrevBids=false;

        },
        (err) => {
            console.log(err);
            this.alertService.add({"type":"danger",
            "message":  "Some error occured, please contact Administrator"});
            this.loadingPrevBids=false;
            this.appInsights.instance.trackException(err);
        });
    this.loadingPrevBids=false;
```

```
        this.appInsights.instance.trackEvent({name:"LoadedPreviousBids"});
```

    }

   This method calls the bid service to get all the bids made on a particular auction and displays them in a table. Finally, to submit a bid, we will create a method named submitBid, which is shown in Listing 4-30.

LISTING 4-30   Submit Bid method

```
public submitBid(){
        this.loadMakeOffer =true;
        this.bidDetail.userId = this.storage.get("userId");
        this.bidDetail.userName=this.storage.get("userName");
        this.httpClient.post(environment.bidAPI+ "/bid?bidAmount="+this.bidDetail.
        bidAmount+"&userID="+this.bidDetail.userId+"&auctionID="+this.auction.
        idAuction+"&userName="+this.bidDetail.userName, this.auction).subscribe(
            (res) => {
                this.alertService.add({"type":"success", "message": "Bid is made
                successfully"});
                this.loadPreviousBids();
                this.loadMakeOffer=false;
            },
            (err) => {
                console.log(err);
                this.alertService.add({"type":"danger", "message":
                "Some error occured, please contact Administrator"});
                this.loadMakeOffer=false;
                this.appInsights.instance.trackException(err);
            });
        this.appInsights.instance.trackEvent({name:"SubmittedBid"});
    }
```

   In Listing 4-30, we are getting the user ID and username from session storage and are then making an HTTP POST request to the bid service. We then are passing bid information as query parameters within the request itself. For a POST request, we subscribed promises for the success and error events and to show the respective alert message on the page.

> **NOTE**   **Promises**
>
> In JavaScript, promises represent processes that are already running and can be chained with the callback functions.

   Now, let's modify the bid details component HTML page and add the section to submit bid information. For that, we will use the same approach and utilize an ngForm directive to create a form, as shown in Listing 4-31.

LISTING 4-31 Submit Bid form

```
<form #f="ngForm" (ngSubmit)="submitBid()">

        <div class="form-group" *ngIf="!loadMakeOffer">
            <label for="bidAmount">Enter Bid Amount (in USD)</label>
            <input [(ngModel)]="bidDetail.bidAmount"
            type="text" class="form-control" name="bidAmount"
            id="bidAmount" required #bidAmount="ngModel">
          <div [hidden]="bidAmount.valid "
            class="alert alert-danger">
            Bid Amount is required
          </div>
        </div>
        <button type="submit" class="btn btn-primary" [disabled]="!f.valid"
        *ngIf="!loadMakeOffer">Submit Bid </button>
        <div class="d-flex justify-content-center">
         <div class="spinner-border text-secondary"
         *ngIf="loadMakeOffer" style="width: 3rem; height: 3rem;"
role="status" >

              <span class="sr-only">Loading...</span>
          </div>
        </div>
      </form>
```

To show the previous bids made for an auction, we need to create an HTML to display the bid information, as shown in Listing 4-32.

LISTING 4-32 Previous Bids table

```
<table class="table table-hover" *ngIf="!loadingPrevBids">
    <thead>
      <tr>
        <th scope="col">#</th>
        <th scope="col">Amount </th>
        <th scope="col">User</th>
        <th scope="col">Bidded At</th>
      </tr>
    </thead>
    <tbody>
     <tr *ngFor="let bid of bids">
        <th scope="row">{{bid.bidID}}</th>
        <td>{{bid.amount}}</td>
        <td>{{bid.customerName}}</td>
        <td>{{bid.bidAt | date: 'dd/MM/yyyy h:mm a' }}</td>
      </tr>
    </tbody>
</table>
```

In this section, we had a quick introduction to some key building blocks of Angular. We discussed creating an Angular app, implementing security, and enabling authorized access to the pages using Route Guard. Further, we also created an auction form and the active auctions page to view active auctions and place bids from the bid details page.

Summary

In this chapter, we developed the Online Auction System application and discussed the following topics:

- We started with developing a back-end based on microservices.
- We developed an auction service and provisioned Azure MySQL database on Azure using Terraform.
- We developed a bid Service and provisioned Azure Cosmos DB with MongoDB API on Azure using Terraform.
- We developed a payment service and provisioned an Azure SQL database on Azure using Terraform.
- We developed a front-end application and learned some key concepts and building blocks of Angular.
- We implement security in a front-end application.

At this stage, the basic application is developed. In the next chapter, we will learn about the communication protocols and implement a pub/sub model. We will go through the entire end-to-end implementation of setting up the resources for pub/sub communication and develop a listener service in .NET Core to listen for events and provide integration with other services.

Chapter 5

Microservices on containers

In this chapter, you will:

- Understand core concepts around containers
- Learn Docker as a container technology
- Use Azure Container Registry as a containers repository
- Learn the Kubernetes Architecture and use Azure Kubernetes Services for container orchestration
- Learn Azure App Service and deploy a front-end app
- Get an overview of WebJobs and deploy a listener service as a webjob

Container infrastructure is one of the four pillars of the cloud native computing foundation. Containers provide a sandboxed environment for the application and put an end to the phrase "it works on my machine" that is quite popular in our industry. Once the application is containerized, the same container can be ported to any machine that supports the respective container technology and contains all the dependencies within the container artifact, which is known as an image.

In the OAS (Online Auction System), we have a few microservices, a front-end, and a background service. In this chapter, we will containerize all the microservices using Docker and deploy them on the Azure Kubernetes Services for container orchestration. For the front-end, we will leverage Azure App Services, which is a PaaS (Platform as a Service) used to host web applications. For the background service—a listener service—we will use Azure WebJobs.

Containers Overview

In the microservices architecture, each microservice is a separate host process that is built on specific technology, language, or framework and maintains a separate database to save domain-specific information. Because the service is representing one part of the application, it can easily be scaled out to accommodate the respective workloads. For example, if you have a service hosted on a physical or virtual machine and there is a need to scale out, a separate machine needs to be provisioned where you install an operating system, install other frameworks or runtimes as needed by the service, and then set up a webservice to host.

Containers provide virtualization at the operating system level, as opposed to VMs, where you need to create a separate VM, install the respective operating system, and then configure everything from scratch as required by your application. With a VM, you might end up with one service hosted on one VM, which is not a cost-effective solution.

With containers, you can simply build an image and run it as a container instance. All the containers share the same kernel from the host operating system, and at the same time, they have their own base image layer for their respective operating system. The container runtime handles all the communication from the container to the host OS kernel. Essentially, a container is a portable unit of deployment that packs application code and related dependencies into a single unit. For a better understanding of how containers compare to VMs, see Figure 5-1.

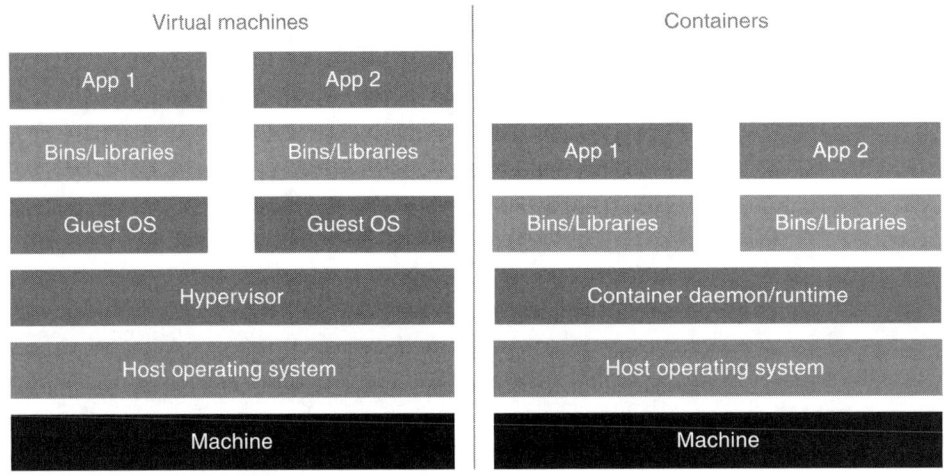

FIGURE 5-1 VMs versus containers

Docker as a container technology

Containers solve the problem of running an application in an isolated sandbox space without having any dependencies on the system. However, for Windows, the base image layer should be specific to the Windows platform only.

Install Docker

To containerize the OAS microservices, you first need to install the Docker tool on your system. Docker can be installed on macOS, Windows, and various Linux distros.

Once Docker is installed, you click the **Docker** icon and open the settings to configure shared drives, CPUs, memory, and disk image location from the tabs on the left, as shown in Figure 5-2.

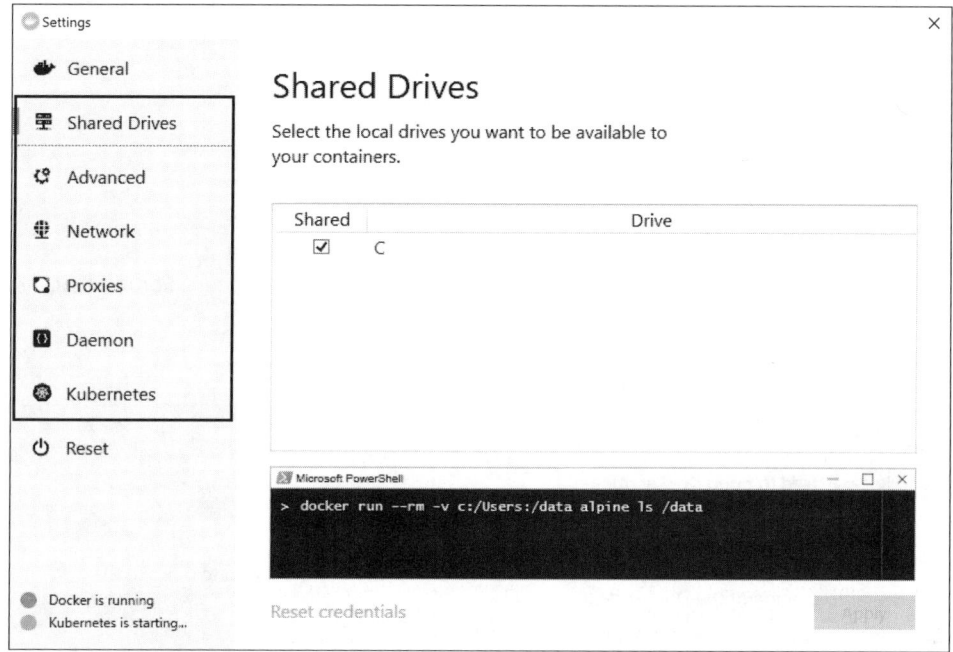

FIGURE 5-2   Docker settings in Windows

To verify that Docker has been configured properly, you can also run the Docker CLI command, as shown in Figure 5-3.

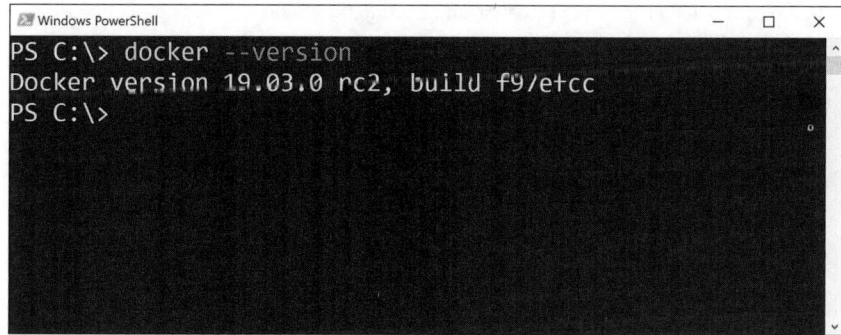

FIGURE 5-3   Running the docker version command

Docker components

Before starting to work with Docker, it is essential to understand its key components. Table 5-1 shows a description of the Docker components.

TABLE 5-1 Docker components

Docker Component	Description
Host	Physical or virtual machine where the Docker daemon/engine is installed
Docker client	Application that interacts with Docker
Docker daemon	Docker engine that takes command from the client application to build a Docker image, run Docker containers, and so on
Docker container	Running instance of the Docker image
Docker registry	Repository of all the Docker images

Docker commands

Docker provides various commands to build, run, and manage your containers. Some of the most widely used commands are shown in Table 5-2.

TABLE 5-2 Docker commands

Docker Command	Syntax
docker build	Used to build Docker images. **Syntax:** `docker build -t imagename:versionno` **Example:** `docker build -t auctionservice:1.0 .` The -t switch is used to tag an image. The dot after version is used to point to the location of the Dockerfile.
docker create	To create a Docker container from the image. **Syntax:** `docker create imagename:versionno` **Example:** `docker create auctionservice:1.0`
docker start	To run the stopped container. **Syntax:** `docker start containerid` **Example:** `docker start f0c78ba19b96`
docker run	To run the container. Instead of creating and starting the container, this single command can create and start the container. **Syntax:** `docker run -d -p hostport:containerport imagename:versionno` **Example:** `docker run -d -p 8080:3000 auctionservice:1.0` The -d switch is used to start the container in a detachable mode. So, when the command is run with the -d switch, the container will be started in a separate thread, and you get the cursor on the same terminal to execute other commands. The -p switch is to map the host port to the container port. The container is running on its own port, but to access it from the local IP, we need to provide the host port.
docker ps	To list the running instances of docker containers.
docker ps -a	To list all the containers that are running or exited because of some error.

Docker Command	Syntax
docker tag	To tag a Docker image to some other name. This is used when uploading the docker image to some container registry. **Syntax:** `docker tag imagename:versionno registryname/imagename:versionno` **Example:** `docker tag auctionservice:1.0 oasregistry.azurecr.io/auctionservice:1.0` The oasregistry.azurecr.io is the Azure Container Registry.
docker push	To push the image to the container registry. **Syntax:** `docker push registryname/imagename:versionno` **Example:** `docker push oasregistry.azurecr.io/auctionservice:1.0`
docker pull	To download the image from the container registry. **Syntax:** `docker pull registryname/imagename:versionno` **Example:** `docker pull mcr.microsoft.com/windows/nanoserver:2004`

NEED MORE INFO? **Other Docker Commands**

Other commands are available at *https://docs.docker.com/engine/reference/commandline/docker*

Linux versus Windows containers

Docker supports both Linux and Windows Containers. The underlying architecture of Linux containers is similar to Windows containers where the Docker engine is taking all the requests from the Docker client and communicating them to the shared kernel of the host OS to use resources (see Figure 5-4).

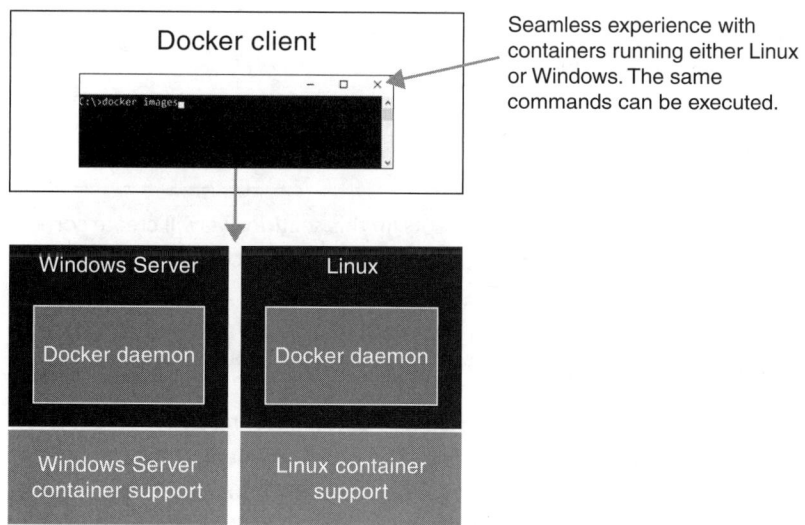

Seamless experience with containers running either Linux or Windows. The same commands can be executed.

FIGURE 5-4 The Docker client uses the same commands for both Windows and Linux

In Windows, Hyper-V (Hypervisor) containers provide isolation that is similar to a VM. Figure 5-5 depicts the difference between Windows and Windows Hyper-V containers.

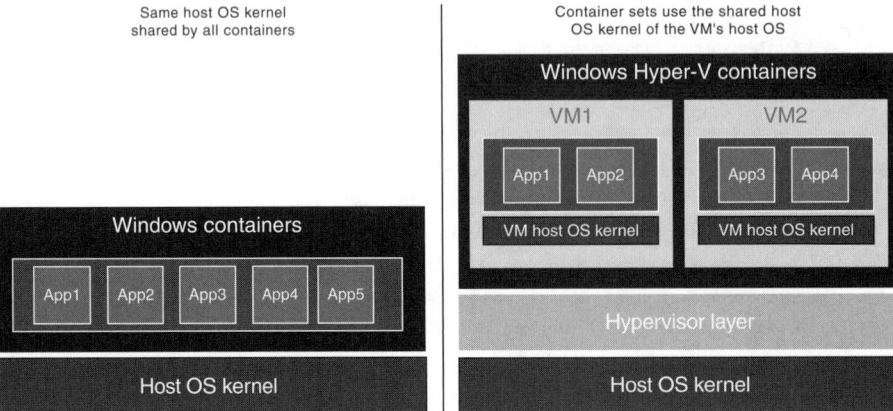

FIGURE 5-5   Windows and Windows Hyper-V containers

Figure 5-5 shows the two modes provided for Windows containers. Windows containers offer process isolation, whereas the Windows Hyper-V containers provide Hyper-V isolation, though both operate similarly.

By default, Windows containers created using the docker run command use process isolation mode. However, you can explicitly specify the --isolation=process switch when running the docker run command to run with process isolation.

```
docker run nanoserver:1809 --isolation=process
```

For Hyper-V isolation, you need to specify the --isolation=hyperv switch to run the container with Hyper-V isolation.

```
docker run nanoserver:1809 --isolation=hyperv
```

Build Docker images

So far, we have discussed containers and Docker concepts. In this section, we will create containers for the OAS microservices, and we will walk through the steps to configure, build, and run Docker containers.

Containerize the auction service

To containerize any service or application using Docker, we need to first create a Dockerfile. A Dockerfile is a text document that contains all the commands or statements needed to build the image. A container is like a fresh machine. To run your application, you need to build the layers to allow your application run. For example, to run a Node.js application, we need to first set up the OS and install Node dependencies, application dependencies, and other dependencies used by the application as prerequisites.

To start building a Docker image for the auction service, let's create a new Dockerfile in the root folder of the auction service workspace. Here is the code snippet for the Auction Service Dockerfile:

```
FROM node:8.9-alpine
WORKDIR /src
COPY package*.json ./
RUN npm install
COPY . . /
EXPOSE 3000
CMD [ "npm", "start" ]
```

The code snippet above is broken down here:

- The first command is the FROM statement, which is used to pull the base image for the container. There are lots of base images available at *https://hub.docker.com* that can be used to build the Docker image. This image will install the base OS files from the Alpine Linux distro and the node dependencies needed to run the Node.js application.

- The second command is WORKDIR, which is used to set the folder directory on the container. In our case, the folder directory is /src.

- The COPY command is used to copy the package.json file to the root folder of the working directory, which is /src. The package.json file contains all the application dependencies and dev dependencies used to build and run the Node.js application.

- The RUN Command is used to install the dependencies, and once the dependencies are downloaded and installed, we copy all the files from the local root folder where the Dockerfile resides into the container's working directory.

- Finally, the EXPOSE command is used to register a port in a container where the application will listen. Use the CMD command to start the application.

To build the image, we need to run the docker build command, as shown in Figure 5-6.

```
docker build -t auctionservice:1.0 .
```

FIGURE 5-6 The docker build command being executed in the PowerShell terminal window

This command looks for the Dockerfile in the root folder and will start executing the steps as defined in the Dockerfile. Once the image is successfully built, verify that the image has been created by running the following command:

```
docker images
```

You will be able to see the image name in the **REPOSITORY** column, the version in the **TAG** column, and the ID of the image in the **IMAGE ID** column, as shown in Figure 5-7.

FIGURE 5-7 The REPOSITORY, TAG, and IMAGE ID columns

So far, the image is built, but the container is not yet created. To create the container, run the docker run command that spins up the new container instance for the auctionservice image.

```
docker run --name auctionservice -d -p 3001:3000 auctionservice:1.0
```

The above command creates and runs the new container instance. The --name switch is used to provide the name of the container. This is optional and if not specified, a random name will be picked and assigned by Docker. The -d switch is used to start in detachable mode, and the -p switch is used to map the host OS port to the container port. We specified 3000 for the container port because the application is configured to run on port 3000. The name has been set to auctionservice, and the version number of the image is the one we built in the previous section (auctionservice:1.0). Once the container is run, we can verify it by running the following command:

```
docker ps
```

Figure 5-8 shows a list of all the running instances in the system.

```
CONTAINER ID        IMAGE                COMMAND
7c1196857e84        auctionservice:1.0   "npm start"
```

FIGURE 5-8 The list of running Docker containers on the system

Finally, to test the auction service, you can navigate to the *http://localhost:3001* and access the web methods exposed by the service.

Containerize the bid service

In this section, we will containerize the bid service, which is built on the JavaSpring Boot framework. First, we will create a Dockerfile in the root folder of the bid service project. Here is the Dockerfile snippet to build the image for the bid service.

```
FROM openjdk:8-jdk-alpine
VOLUME /tmp
ARG DEPENDENCY=target/dependency
```

```
COPY ${DEPENDENCY}/BOOT-INF/lib /app/lib
COPY ${DEPENDENCY}/META-INF /app/META-INF
COPY ${DEPENDENCY}/BOOT-INF/classes /app
ENTRYPOINT ["java","-cp","app:app/lib/*","com.onlineauction.bidservice.bidservice.
DemoApplication"]
```

The code snippet above is broken down here:

- The Docker file starts by importing the image from `openjdk:8-jdk-alphine`.

- It has a `ARG DEPENDENCY` parameter, which points to the directory that contains the unpacked .jar file.

- The unpacked .jar contains folders, such as `BOOT-INF/lib` and `BOOT-INF/classes`, which contain the jars and application classes within it.

- `ENTRYPOINT` denotes the actual entry point class, `DemoApplication`, which will be run on startup.

The Docker image for JavaSpring Boot application can be built in the same way by running the `docker build` command. However, JavaSpring Boot also provides a wrapper to build Docker images using Maven commands. So, in this section, we will build a Docker image using the `mvnw` command.

To start building the Docker image for the bid service using the `mvnw` command, we need to add following dependencies in our bid service project. Modify the `pom.xml` file and add the packages shown in Listing 5-1:

LISTING 5-1   Adding the Dockerfile Maven package

```
<properties>
    <docker.image.prefix>springio</docker.image.prefix>
</properties>
<build>
    <plugins>
        <plugin>
            <groupId>com.spotify</groupId>
            <artifactId>dockerfile-maven-plugin</artifactId>
            <version>1.4.9</version>
            <configuration>
                <repository>${docker.image.prefix}/${project.artifactId}</repository>
            </configuration>
        </plugin>
    </plugins>
</build>
```

We will add the configuration shown in Listing 5-2 for the dependency plugin to ensure the jar is unpacked before the Docker image is created.

LISTING 5-2   Adding the Maven package configuration

```
<plugin>
    <groupId>org.apache.maven.plugins</groupId>
    <artifactId>maven-dependency-plugin</artifactId>
    <executions>
        <execution>
            <id>unpack</id>
            <phase>package</phase>
            <goals>
                <goal>unpack</goal>
            </goals>
            <configuration>
                <artifactItems>
                    <artifactItem>
                        <groupId>${project.groupId}</groupId>
                        <artifactId>${project.artifactId}</artifactId>
                        <version>${project.version}</version>
                    </artifactItem>
                </artifactItems>
            </configuration>
        </execution>
    </executions>
</plugin>
```

To build the image, use the following command:

```
mvnw install dockerfile:build
```

You can find the mvnw command utility in the root folder of the bid service. The important thing to consider here is that you must enable the **Expose Daemon On tcp://localhost:2375 Without TLS** checkbox in the Docker settings because, by default, the plugin tries to connect to Docker on port 2375. After running the mvnw install dockerfile:build command, you will see the steps shown in Figure 5-9 in the VS Code terminal window.

FIGURE 5-9   The creation of a Docker image using the mvnw command

Once the image is successfully built, you can verify that the Docker image has been properly built, as shown in Figure 5-10.

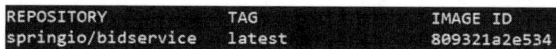

REPOSITORY	TAG	IMAGE ID
springio/bidservice	latest	809321a2e534

FIGURE 5-10 The bid service Docker image is properly built

springio is added as a prefix for the image, and it is specified in the following section of the pom.xml file.

```
<properties>
    <java.version>1.8</java.version>
    <docker.image.prefix>springio</docker.image.prefix>
</properties>
```

However, we can also modify it to use any prefix we want. Finally, the bid service can run using the same docker run command, as shown below, which spins up the new container for the bid service.

```
docker run -p 9090:9090 -t springio/bidservice:latest
```

Containerize the payment service

Now, we will configure the Dockerfile for the payment service that is built on .NET Core. First, we need to create a new Dockerfile in the root folder of the payment service project. Listing 5-3 shows the Dockerfile for the payment service.

LISTING 5-3 Payment service Dockerfile

```
FROM mcr.microsoft.com/dotnet/core/aspnet:3.0-buster-slim AS base
WORKDIR /app
EXPOSE 80
EXPOSE 443

FROM mcr.microsoft.com/dotnet/core/sdk:3.0-buster AS build
WORKDIR /src
COPY ["PaymentService/PaymentService.csproj", "PaymentService/"]
RUN dotnet restore "PaymentService/PaymentService.csproj"
COPY . .
WORKDIR "/src/PaymentService"
RUN dotnet build "PaymentService.csproj" -c Release -o /app/build

FROM build AS publish
RUN dotnet publish "PaymentService.csproj" -c Release -o /app/publish
```

```
FROM base AS final
WORKDIR /app
COPY --from=publish /app/publish .
ENTRYPOINT ["dotnet", "PaymentService.dll"]
```

This is a multi-staged Dockerfile. Multi-stage builds were introduced in Docker version
17.0.5, and they help you reduce the overall image of your application by splitting it into different layers. Each instruction or statement is a layer of the image, and it is a good practice
to clean up the resources once your application is built. Only the release artifacts should be
copied and made part of the final image. In the previous Dockerfile, we used `mcr.microsoft.`
`com/dotnet/core/aspnet:3.0-buster-slim` as the base image. This image needs to run the .NET
Core application. We named this stage `base` and exposed ports **80** and 443. When we create a
new .NET Core, the default port is 5000 for HTTP and 5001 for HTTPS. However, when we build
the Docker image, the base image `mcr.microsoft.com/dotnet/core/aspnet:3.0-buster-slim`
internally sets the `ASPNETCORE_URLS` environment variable to port 80 for HTTP. If you have not
set the `app.UseUrl` in `Program.cs`, then the application will listen on port 80 for HTTP. You can
also override the default port by specifying `ENV ASPNETCORE_URLS=http://+:5000` or any other
port if required.

To build the payment service, we need to pull another image, `mcr.microsoft.com/dotnet/`
`core/sdk:3.0-buster`, which contains the .NET Core SDK and all the related files to build the
application. We named this stage `build`, set the working directory to `/src`, and copied the
`.csproj` file to the `PaymentService` folder inside `/src`. The `.csproj` file contains all the dependencies; target framework version; and properties, such as authors, company, language,
assembly title, and other configuration attributes, which are used by the .NET Compiler to
restore the dependencies using the `dotnet restore` command.

After restoring all the dependencies from `Nuget.org`, we copied all the files from the current
folder to the container folder, which in our case is `/src`. We then ran the `dotnet build` command to build the application where `-c Release` is used to build the application in release
mode. The `-o` switch is used to mention the output directory where all the build artifacts will be
generated.

To publish artifacts, we defined another layer from the build and published the artifacts to
the `/app/publish` folder using the `dotnet publish` command.

We used the runtime image and copied the files from the `/app/publish` folder to the root
folder, `/app`, in the final runtime image. Lastly, we specified the `ENTRYPOINT` to spin up the payment service application by using the `dotnet run` command. Build the image by running the
following command:

```
docker build -t paymentservice:1.0
```

Once the image is built, we can run the payment service using the `docker run` command, as
shown below:

```
docker run -p 80:80 -t paymentservice:1.0
```

Deploy images to Azure Kubernetes Services

In the previous section, we built and ran the Docker images for the auction, bid, and payment services. For a production scenario, we need to deploy them to a container orchestration platform that allows us to easily manage containers in terms of scalability and availability.

When we talk about the container orchestration platform, Kubernetes is one of the most widely used. In this section, we will first understand the overall architecture of Kubernetes, how to provision and use the managed Kubernetes on Azure known as AKS (Azure Kubernetes Services), and we will explore the steps needed to deploy the Docker images from the local machine to a private registry; in our case, we will use the Azure Container Registry. Finally, we will deploy to AKS.

Kubernetes architecture

Kubernetes is a cluster of VM (Virtual Machines) or physical machines known as nodes in Kubernetes terms. There are worker nodes and a master node that manages the worker nodes. At a high level, the master node consists of a control plane, a storage system known as ETCD, and a scheduler to schedule pods. Whereas each worker node contains the container runtime, Kubelet is an agent used to communicate with the master node and pods where one or more than one container runs inside it. Figure 5-11 shows the schematic representation of how a Kubernetes cluster looks.

Figure 5-11 represents the Kubernetes cluster architecture, which contains two worker nodes and a master node to manage them. The deployment starts when a user declaratively creates the YAML file for specific deployment in Kubernetes, such as pod, deployment, service and so on. The user runs the `kubectl apply` command, which points to the full path of the YAML file containing the metadata and specifications for the kind of deployment that needs to be done. Here is a sample command to apply the Kubernetes deployment:

```
kubectl apply -f k8deploy.yaml
```

The `kubectl apply` command communicates to the API Server part of the master node and passes the content defined in the YAML file in JSON format over the HTTP endpoint. The API server is the Kubernetes control plane frontend, and all the commands executed using `kubectl` communicate to the API Server. The API Server saves the configuration in the ETCD store in key/value pairs. The API Server then talks to other components known as the scheduler and controller manager to deploy the containers to different nodes inside pods.

The scheduler always watches for pods that are not assigned to any pods and deploys them in available nodes. The controller manager is responsible for overseeing the nodes and responding when any node goes down. The controller manager also compares the current state with the desired state of any deployment by referencing the ETCD store and maintaining the number of pods as desired by a specific deployment. It also helps establish a connection between pods and services. Lastly, for new namespaces, it also creates accounts and respective API access tokens.

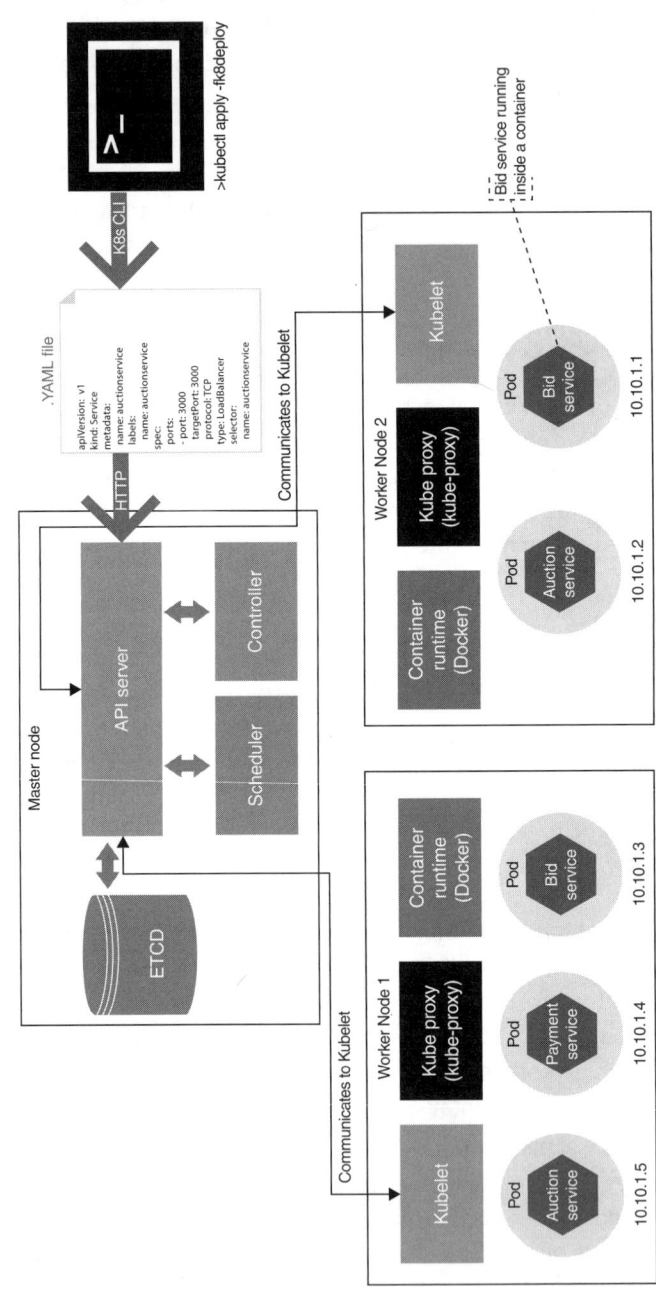

FIGURE 5-11 The high-level architecture of Kubernetes

On the worker node side, the container runtime is the actual runtime that manages the containers within the node itself. For example, when building and running Docker containers, we need the Docker runtime to be installed on the local machine. The container runtime is required to run Docker containers.

The Kubelet is the agent running on each node that ensures the containers are running inside a Pod, whereas the Kube proxy is a network proxy that conserves network rules on nodes.

Provision Azure Kubernetes Services

AKS (Azure Kubernetes Service) is the managed Kubernetes (K8s) service on Azure. AKS is a serverless Kubernetes that can be provisioned in Azure in minutes. You can orchestrate containers and run Kubernetes without any management overhead. It enables elastic provisioning that provides an ability to add event-driven autoscaling and triggers through KEDA.

> **NOTE More about KEDA**
>
> KEDA is the event-driven autoscaling component that allows users to set rules to auto-scale an Kubernetes cluster. To learn more about KEDA, see *https://github.com/kedacore/keda*

In AKS, you get an SLA of 99.95 percent high availability, and you pay only for the virtual machines and the associated storage and networking resources consumed.

> **NOTE More about AKS**
>
> To learn more about AKS, see *https://azure.microsoft.com/en-us/services/kubernetes-service/#solution-architectures*

To provision AKS, we will utilize Terraform. In VS Code, open the same Terraform project you created earlier and create a new Terraform file named AKS.tf. Listing 5-4 shows the AKS content configuration:

LISTING 5-4 AKS content configuration

```
provider "azurerm" {
  features {}
}
resource "azurerm_resource_group" "aks" {
  name     = "OSS "
  location = "West Europe"
}

  resource "azurerm_kubernetes_cluster" "aks" {
    name              = "onlineauctionkubernetes"
```

```
location            = azurerm_resource_group.aks.location
resource_group_name = azurerm_resource_group.aks.name
dns_prefix          = "onlineauctionkubernetes"

default_node_pool {
  name       = "default"
  node_count = 3
  vm_size    = "Standard_DS2_v2"
}

identity {
  type = "SystemAssigned"
}

addon_profile {
  aci_connector_linux {
    enabled = false
  }

  azure_policy {
    enabled = false
  }

  http_application_routing {
    enabled = false
  }

  kube_dashboard {
    enabled = true
  }

  oms_agent {
    enabled = false
  }
}
}
```

You can run the deployment from VS Code by first initializing the cloud shell using the `Azure Terraform: Init` command from the command palette. Once the cloud shell is initialized, run the `Azure Terraform: Plan` to validate the deployment and establish a plan to deploy it on Azure. Finally, run the `Azure Terraform: Apply` to apply the planned deployment.

> **NEED MORE INFO?** **More on AKS**
>
> To set up AKS with advanced options using Terraform, see *https://github.com/ terraform-providers/terraform-provider-azurerm/tree/master/examples/kubernetes*

Provision the Azure Container Registry

To deploy containers on AKS or any other orchestration engine, we need to first upload them to the container registry. For the OAS, we leveraged ACR (Azure Container Registry) to push images to its repository, and then we used that image to deploy its container inside an AKS cluster.

The ACR can be provisioned from Azure portal or by running Azure commands using Azure CLI, ARM templates, Terraform, and others. However, we will use Terraform to deploy ACR.

First, create a new Terraform file and name it `ACR.tf`. Here is the content of the Terraform file to provision ACR:

```
resource "azurerm_resource_group" "rg" {
  name     = "OSS"
  location = "West Europe"
}
resource "azurerm_container_registry" "acr" {
  name                = "onlineauctionregistry"
  resource_group_name = azurerm_resource_group.rg.name
  location            = azurerm_resource_group.rg.location
  sku                 = "Standard"
  admin_enabled  = true
}
```

The above `ACR.tf` file can be applied using VS Code and running the commands (`Init`, `Plan`, and `Apply`) in sequence, as we did in the previous section when we set up AKS. The ACR will be created with the name `onlineauctionregistry` inside the OSS resource group, and the location will be West Europe. The `admin_enabled` bit is set to `true` so we can connect to ACR to push images. Moreover, these credentials will also be used by the AKS to pull images from ACR.

Push services to ACR

To deploy the auction service to AKS, we need to first tag the Docker image to the same container registry instance created in the previous section. In our case, the ACR is onlineauction-registry.azurecr.io. We can tag the image using the `docker tag` command. Open PowerShell or any other terminal and run the following command:

```
docker tag auctionservice:1.0 onlineauctionregistry.azurecr.io/auctionservice:1.0
```

Once the image is tagged, we need to connect to the ACR using the following syntax. First log in to Azure using Azure CLI. Then log in to the Azure portal using Azure CLI and execute the following command:

```
az login
```

The `az login` command will open the browser and ask you to sign in with your Azure credentials. Once the log in is successful, you can see the list of subscriptions on the console window. Next, log in to the container registry by running the following command:

```
az acr login –name onlineauctionregistry
```

Once the log in is successful, the tagged Docker image can be pushed by running the command below:

```
docker push onlineauctionregistry.azurecr.io/auctionservice:1.0
```

Finally, the image is pushed, and you can verify it by logging in to the Azure portal and accessing the **Repositories** tab in ACR. To push the bid and payment services, we need to repeat the same steps. Here are the commands to tag the bid and payment services:

```
docker tag bidservice:1.0 onlineauctionregistry.azurecr.io/bidservice:1.0
docker tag paymentservice:1.0 onlineauctionregistry.azurecr.io/paymentservice:1.0
```

And here are the commands to push them into ACR:

```
docker push onlineauctionregistry.azurecr.io/bidservice:1.0
docker push onlineauctionregistry.azurecr.io/paymentservice:1.0
```

Figure 5-12 shows the Docker images for all the services pushed to ACR.

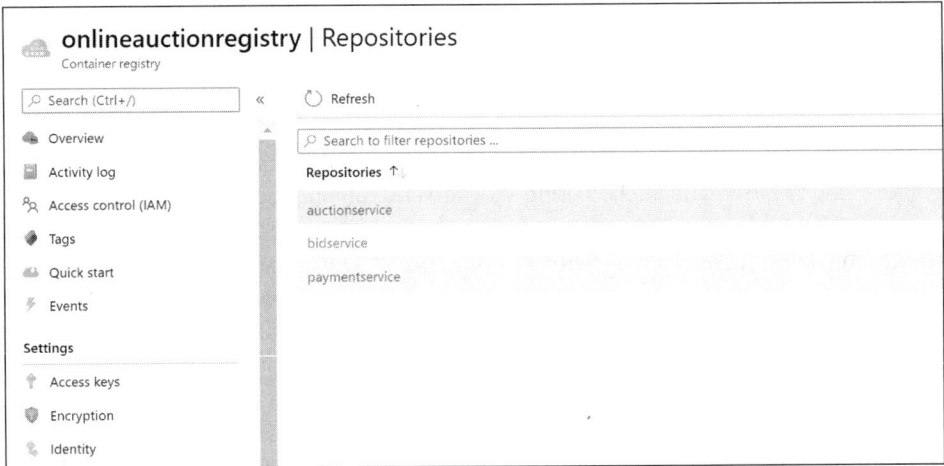

FIGURE 5-12   The published Docker images inside Azure Container Registry

Deploy services to AKS

To deploy services in AKS, we need to run `kubectl` commands. To access the recently created AKS cluster in the previous section, we need to first get its credentials. This is the command to get the credentials for AKS:

```
az aks get-credentials --name NAMEOFAKS --resource-group RESOURCEGROUPNAME
```

In our case, the command will look like this:

```
az aks get-credentials --name onlineauctionkubernetes --resource-group OSS
```

By executing the above command, the credentials are downloaded, and the Kubernetes CLI configured to use them to trigger deployments and manage the AKS cluster.

Next, we need to create a secret that we can refer to when configuring the deployment file to create pods in AKS. This secret will be used by AKS to download the images from ACR and deploy them inside pods. To create a secret, we need to use Kubernetes CLI and run the following command:

```
kubectl create secret docker-registry SECRET_NAME
--docker-server=REGISTRY_NAME.azurecr.io
--docker-email=EMAIL
--docker-username=SERVICE_PRINCIPAL_ID
--docker-password=PASSWORD
```

The code above is broken down here:

- The SECRET_NAME can be any name.
- The REGISTRY_NAME should be the fully qualified name of the container registry.
- The EMAIL can be any email address.
- The SERVICE_PRINCIPAL_ID should be the username of the container registry.
- The PASSWORD is the registry password.

To get the username and password from ACR, go to the Azure portal and access the ACR resource. Select **Access Keys** on the **Settings** panel and copy the username and password to use when creating the secret. Replace the following values as per your configuration and run this code on your system:

```
kubectl create secret docker-registry oasregistrysecret
--docker-server=onlineauctionregistry.azurecr.io
--docker-email=oas@oas.com
--docker-username=onlineauctionregistry
--docker-password=123456789
```

Create a deployment object for OAS microservices

Kubernetes provides both imperative and declarative ways of creating objects in a Kubernetes cluster. The imperative way is to run the `kubectl` commands and pass the parameters inline when running the command itself. With the declarative approach, we need to create a YAML file and configure the properties as required by the template type of that object.

In this section, we will create deployments for all three services. Each object (pod, deployment, service, container, and others) have their own template files. However, the root properties are the same and include the `apiVersion`, `kind`, `metadata`, and `spec`.

Listing 5-5 shows the deployment template for the auction service that creates a deployment object in Kubernetes and creates pods.

```
apiVersion: apps/v1
kind: Deployment
metadata:
  labels:
    name: auctionservice
  name: auctionservice
spec:
  replicas: 3
  selector:
    matchLabels:
      name: auctionservice
  template:
    metadata:
      labels:
        name: auctionservice
    spec:
      containers:
      - image: onlineauctionregistry.azurecr.io/auctionservice:v1
        name: auctionservice
        ports:
          - containerPort: 3000
      imagePullSecrets:
      - name: oasregistrysecret
```

The apiVersion is the version of the template, whereas the kind denotes the kind of object we are creating. The metadata contains the name of the object (that is deployment) and some labels if needed. The spec section holds the settings of the deployment. The replicas specify the number of pods we want to create for an auction service. The section after template holds the pod-related configuration.

Each object holds two sections: metadata and spec. The metadata property inside the template contains the metadata information about the pod, and the spec property inside the template holds the pod specification. The pod specification contains the container information such as the fully qualified image name, name of the pod, ports, and imagePullSecret. The imagePullSecret should contain the exact value of the secret created earlier, which is used by the AKS to pull the image from the container registry—in our case, ACR.

The selector property in the deployment spec section is used to link the deployment object with the pod object. matchLabels holds the exact key/value pair as specified in the labels section of the pod specification. Figure 5-13 shows the YAML file for the K8s Deployment object and the association between labels and selectors.

```
apiVersion: apps/v1
kind: Deployment
metadata:
  labels:
    name: auctionservice
  name: auctionservice
spec:
  replicas: 3
  selector:
    matchLabels:
      name: auctionservice
  template:
    metadata:
      labels:
        name: auctionservice
    spec:
      containers:
      - image: onlineauctionregistry.azurecr.io/auctionservice:v1
        name: auctionservice
        ports:
          - containerPort: 3000
      imagePullSecrets:
      - name: oasregistrysecret
```

Labels and selectors used to establish an association between deployment and pod

FIGURE 5-13 YAML file for Kubernetes Deployment object

In Kubernetes, there is a separate template for pod objects, though when we use deployment kind, we can specify the replicas as well, which ensures that the desired number of pods will be kept running inside the Kubernetes cluster. For example, the following code snippet shows the kind as Pod, which just deploys an auction service container inside a single pod. No replicas property is available to define the number of pods we want.

```
apiVersion: v1
kind: Pod
metadata:
  name: auctionservice
spec:
  containers:
  - name: auctionservice
    image: onlineauctionregistry.azurecr.io/auctionservice:1.0
  imagePullSecrets:
  - name: oasregistrysecret
```

To create this deployment on AKS, we need to first save this file (save it as AuctionService K8deploy.yaml) and then run the following command using Kubernetes CLI:

```
Kubectl apply -f AuctionServiceK8deploy.yaml
```

You can run the same command for the bid and payment services. Listing 5-6 shows the configuration for the bid service:

```
apiVersion: apps/v1
kind: Deployment
metadata:
  labels:
    name: bidservice
  name: bidservice
spec:
  replicas: 3
  selector:
    matchLabels:
      name: bidservice
  template:
    metadata:
      labels:
        name: bidservice
    spec:
      containers:
      - image: onlineauctionregistry.azurecr.io/bidservice:1.0
        name: bidservice
        ports:
          - containerPort: 9090
      imagePullSecrets:
      - name: oasregistrysecret
```

Listing 5-7 shows the configuration for the payment service:

LISTING 5-7 Kubernetes deployment object template configuration for the payment service

```
apiVersion: apps/v1
kind: Deployment
metadata:
  labels:
    name: paymentservice
  name: paymentservice
spec:
  replicas: 3
  selector:
    matchLabels:
      name: paymentservice
  template:
    metadata:
      labels:
        name: paymentservice
    spec:
      containers:
      - image: onlineauctionregistry.azurecr.io/paymentservice:v1
```

```
        name: paymentservice
      ports:
        - containerPort: 80
      imagePullSecrets:
      - name: oasregistrysecret
```

Once the deployment is done, we can verify it by running the `kubectl get deploy` command, as shown in Figure 5-14.

```
Windows PowerShell
PS C:\> kubectl get deploy
NAME             DESIRED   CURRENT   UP-TO-DATE   AVAILABLE   AGE
auctionservice   3         3         3            3           406d
bidservice       3         3         3            3           301d
paymentservice   3         3         3            3           302d
PS C:\>
```

FIGURE 5-14 Deployment objects inside an AKS cluster

The pods can be verified by running the `kubectl get po` command, as shown in Figure 5-15.

```
Windows PowerShell
PS C:\> kubectl get po
NAME                            READY   STATUS    RESTARTS   AGE
auctionservice-67bddd7955-bl8cb 1/1     Running   0          175d
auctionservice-67bddd7955-c9xgx 1/1     Running   0          175d
auctionservice-67bddd7955-zzcsx 1/1     Running   0          175d
bidservice-655f76584b-7z7w7     1/1     Running   0          168d
bidservice-655f76584b-jsz4z     1/1     Running   0          168d
bidservice-655f76584b-zlg5v     1/1     Running   0          168d
paymentservice-848f69c66d-8vs2x 1/1     Running   0          236d
paymentservice-848f69c66d-9xc56 1/1     Running   0          236d
paymentservice-848f69c66d-bckxn 1/1     Running   0          236d
PS C:\>
```

FIGURE 5-15 The pods objects running inside an AKS cluster

Create a service object for OAS microservices

So far, we have created the deployments, but we cannot access the services outside the Kubernetes cluster. To do so, we need to create a `kind: Service` object. With a `Service` object, we can get the public IP address through which we can access the microservices or containers running inside pods. If we have more than one pod where a service is running, the `Service` object acts as a Load Balancer, depending on the `Service` configuration, and automatically routes the request to the available pod. Listing 5-8 shows the `Service` object template configuration for the auction service:

```
apiVersion: v1
kind: Service
metadata:
  name: auctionservice
  labels:
    name: auctionservice
spec:
  ports:
  - port: 3000
    targetPort: 3000
    protocol: TCP
  type: LoadBalancer
  selector:
    name: auctionservice
```

In Listing 5-8, the `kind` is Service, the `metadata` holds the Service `name` and `labels`. The `spec` contains service-specific configuration properties, such as the `port` and `targetPort`. The `port` is the service port, whereas the `targetPort` should be equivalent to the port you have defined in the deployment object template created in the previous section. The `type` is set to Load Balancer, so it can act as a load balancer and route traffic to the available pod. Finally, the `selector` is used to enable association with the deployment.

Save this file and name it AuctionServiceK8Service.yaml. Run it using this Kubernetes CLI command:

```
kubectl apply -f AuctionServiceK8Service.yaml
```

For the bid and payment services, we will create separate files and do the configuration individually. For the bid service, Listing 5-9 shows the content for the `kind: Service` object.

LISTING 5-9   Kubernetes service object template configuration for bid service

```
apiVersion: v1
kind: Service
metadata:
  name: bidservice
  labels:
    name: bidservice
spec:
  ports:
  - port: 9090
    targetPort: 9090
    protocol: TCP
  type: LoadBalancer
  selector:
    name: bidservice
```

Lastly, Listing 5-10 shows the `Service` kind configuration for payment service:

LISTING 5-10 Kubernetes service object template configuration for the payment service

```
apiVersion: v1
kind: Service
metadata:
  name: paymentservice
  labels:
    name: paymentservice
spec:
  ports:
  - port: 80
    targetPort: 80
    protocol: TCP
  type: LoadBalancer
  selector:
    name: paymentservice
```

After deploying all the services, we can verify it by running the `kubectl get services` command, as shown in Figure 5-16.

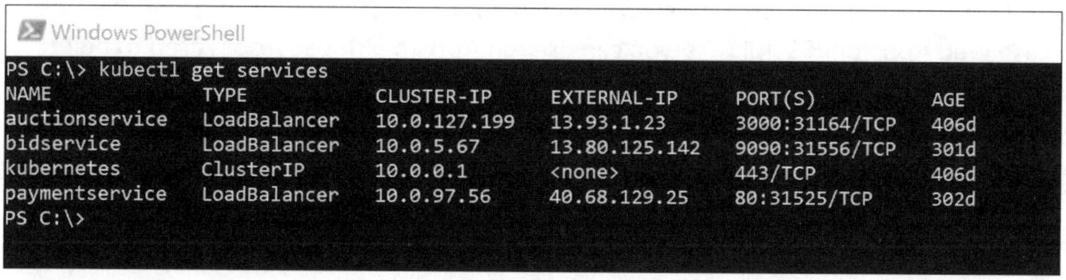

FIGURE 5-16 The services deployed inside an AKS cluster

The services can be accessed using the public IP address specified in the `EXTERNAL-IP` column.

Deploy a front-end application in the Azure App Service

Azure App Service is a managed PaaS (Platform as a Service) offering in Azure that helps developers focus on the application code instead of concentrating on where the website will be hosted, configuring a web server, or setting up a website.

The Azure App Service enables a quick deployment model to deploy a web application that runs inside a container or on a VM with no maintenance and easy scalability options. We can deploy applications built on .NET, .NET Core, Java, Node.js, PHP, Python, and Ruby. Moreover, you can also connect to on-premises resources for the websites hosted as Azure App Services using virtual networks or by setting up the dedicated and isolated App Service Environment. Figure 5-17 shows the types of applications supported in the Azure App Service.

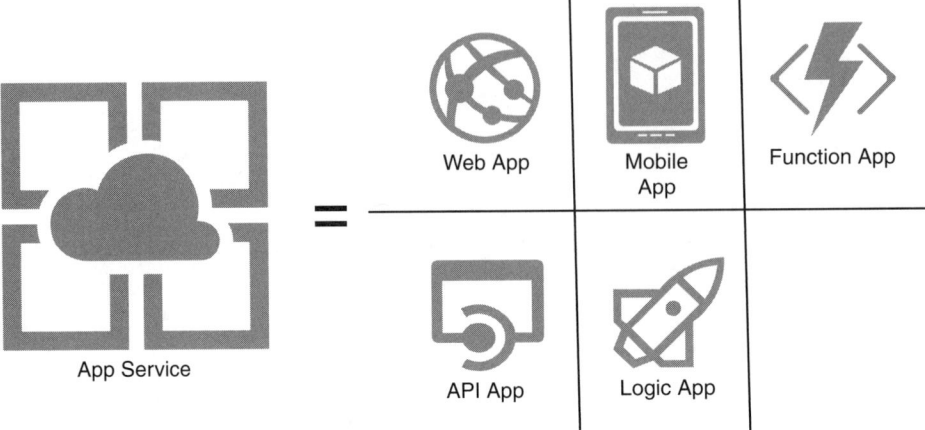

FIGURE 5-17   Azure App Service application types

Azure App Service allows you to host four kind of applications. For front-end applications, we can use Web App; for RESTful APIs, you can use API App; the Mobile App is used to build a mobile app with a fully connected backend API; and the Logic App is used to build codeless configuration-driven workflows. Lastly, the Function App provides a serverless, event-driven experience to accelerate your development.

> **NEED MORE INFO?**   To learn more about Azure App Service, see *https://docs.microsoft. com/en-us/azure/app-service*

We will use Web App of an Azure App Service to deploy our front-end application built on Angular. First, we need to create a new Terraform file and named it AppService.tf.

Listing 5-11 shows the Terraform snippet to configure Azure App Service.

LISTING 5-11   Configure Azure App Service with Terraform

```
resource "azurerm_app_service_plan" "appserviceplan" {
  name                = "oasappserviceplan"
  location            = "West Europe"
  resource_group_name = "OSS"
  sku {
    tier = "Standard"
    size = "S1"
  }
}
```

```
resource "azurerm_app_service" "appservice" {
  name                 = "onlineauctionapp"
  location             = "West Europe"
  resource_group_name = "OSS"
  app_service_plan_id = azurerm_app_service_plan.appserviceplan.id
}
```

In VS Code, execute the script in Listing 5-11 using the same Terraform commands (Init, Plan, and Apply) in sequence. Once the command is executed successfully, the App Service Plan and App Service are created in Azure. The App Service is an empty website that you can access from an endpoint provided when it is provisioned.

Now, because the App Service has been set up, we can now deploy the OAS web application. To do so, we will first build and create the OAS web application package and then deploy it to the Azure App Service.

Go to the root folder of your Angular application and run the following command to build and create the optimized package for deployment:

```
npm run prod-build-dev
```

Once the command is executed successfully, you will find a package named as onlineauction-app created inside the dist folder. For Angular, we need to add the web.config to define some rewrite rules. Create a new file in the root folder inside the dist/onlineauction-app and add the content in Listing 5-12 to the web.config file:

LISTING 5-12   Adding web.config for Angular application

```
<configuration>
<system.webServer>
    <rewrite>
      <rules>
        <rule name="Main Rule" stopProcessing="true">
              <match url=".*" />
              <conditions logicalGrouping="MatchAll">
                  <add input="{REQUEST_FILENAME}" matchType="IsFile" negate="true" />
                  <add input="{REQUEST_FILENAME}" matchType="IsDirectory" negate="true" />
              </conditions>
              <action type="Rewrite" url="/" />
          </rule>
      </rules>
    </rewrite>
</system.webServer>
</configuration>
```

Now to deploy it to Azure App Service, we can use VS Code. You have to first install the Azure App Service extension in VS Code. This can be installed by going to the Extensions bar in VS Code by pressing Ctrl+Shif+X and searching for the Azure App Service extension by typing in the Extensions Marketplace search box at the left, as shown in Figure 5-18.

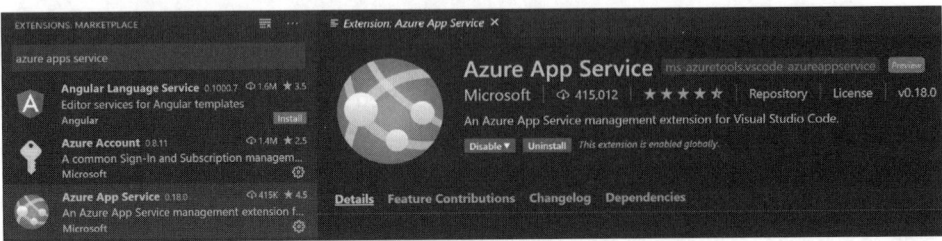

FIGURE 5-18   The Azure App Service extension in VS Code

Install the extension shown in Figure 5-18 and then connect to the Azure subscription, as shown in Figure 5-19.

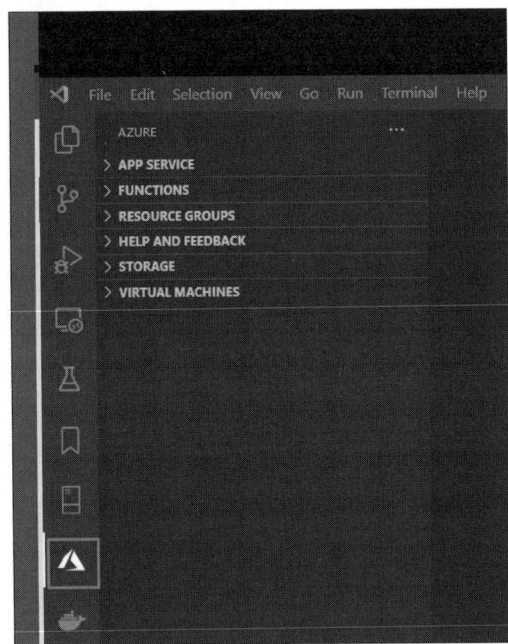

FIGURE 5-19   The Azure panel used in VS Code to access your Azure resources directly from VS Code

Once you are connected, expand the Azure App Service node, and select your service that you have just created, as shown in Figure 5-20.

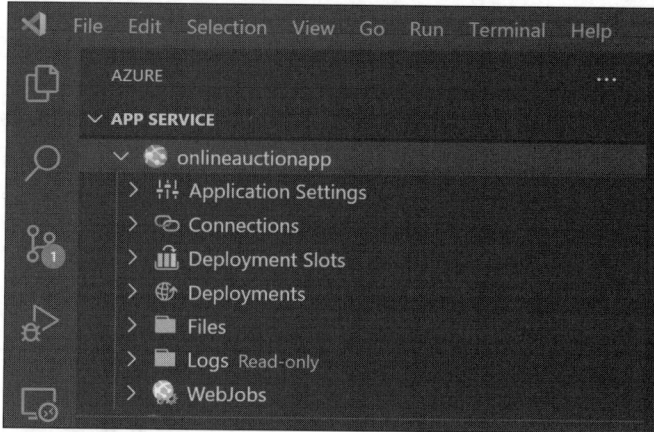

FIGURE 5-20 Selecting the App Service

To deploy the website, right-click the **onlineauctionapp** and select the **Deploy To Web App** option, as shown in Figure 5-21.

FIGURE 5-21 Deploy To Web App

It will ask you to select the folder where the published artifacts reside. Select the same published folder created above and deploy it to your App Service. Once the application is deployed, you can access it from the Azure App Service URL, as shown in Figure 5-22.

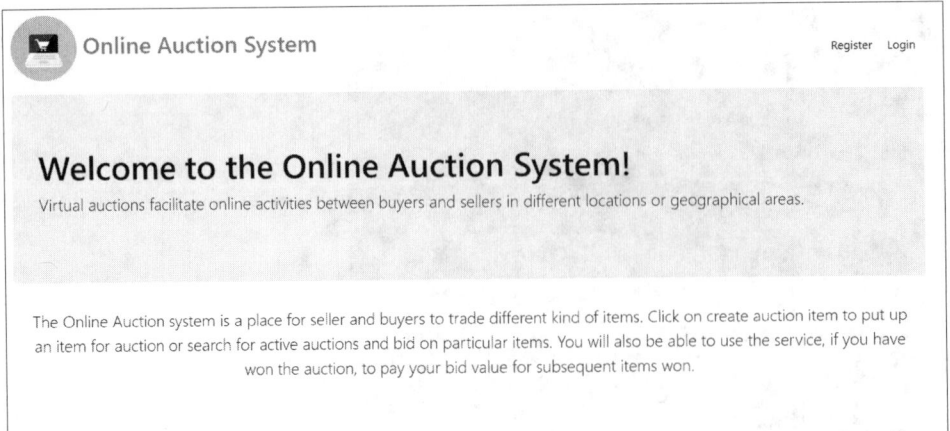

FIGURE 5-22 The Online Auction System website

In this section, we hosted the OAS front-end application built on Angular as an Azure App Service. Let's now deploy our Kafka listener service as an Azure WebJob.

Deploy the Kafka Listener Service as an Azure WebJob

Azure WebJobs are used to run background tasks, which is a feature of the Azure App Service that allows you to run any script or program in the same instance of Azure App Service. To deploy the Kafka listener service inside a WebJob, we need to create a published package and then from the Azure portal, we can deploy that package. Because the Kafka Listener Service is built on .NET Core, we can create a published package by running the following command:

```
dotnet publish -c release
```

This will create a published artifact with release configuration. You can access the folder from the `bin/Release/publish` folder. The published folder must be compressed in order to deploy it to an Azure WebJob.

Compress the folder using WinZip, and then go to the Azure portal and access the same `OnlineAuctionApp` App Service created above. From the Azure App Service, select **WebJobs** from the **Settings** pane and click the **Add** button. Enter the name of the App Service as `KafkaService` and select the path of the `publish.zip` file, as shown in Figure 5-23.

We have defined type as continuous because we don't want this to be triggered from some external source. We have built logic within the Kafka Service itself to continuously listen for Kafka events and process. The **Single Instance** is used because we need to run only one instance of this WebJob to run in the background.

Finally, the WebJob is deployed, and we have completed the deployment of all the microservices, the front-end application, and the background service on Azure.

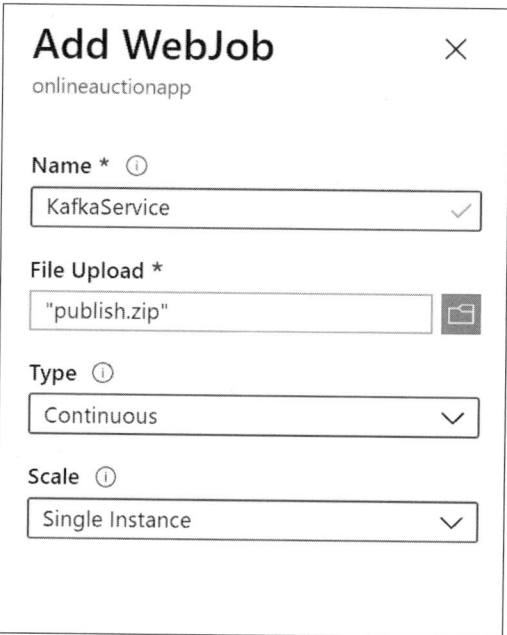

FIGURE 5-23 The configuration needed to deploy the Kafka Listener Service as a WebJob

Summary

In this chapter, we learned the following topics:

- We began with a brief introduction to containers and learned about Docker as a container technology.

- We set up Docker on a local machine and then built and deployed images locally.

- We then explored Kubernetes and learned about the Kubernetes architecture and what AKS offers.

- We set up Azure Container Registry and tagged and pushed images to it and then deployed images to AKS.

- For front-end applications, we chose Azure App Service and deployed the OAS front-end using Azure Web App.

- Finally, we used Azure WebJobs to deploy the Kafka listener service and learned the steps that must be executed to publish and deploy this service as an Azure WebJob.

In the next chapter, we will learn some patterns and practices used for communications between microservices and implement pub/sub messaging using Azure Event Hubs with the Kafka protocol.

Chapter 6

Communication patterns

In this chapter, you will:

- Learn different communication patterns in microservices
- Learn about pub/sub technologies and those used in the Online Auction System (OAS)
- Go through the entire end-to-end process of setting up communication for the OAS

Early in Chapter 1, we mentioned that microservices can sometimes be difficult to work with because of the plethora of communication patterns that one must know and use in the architecture. In this chapter, we want to hedge that concern by walking you through several communication patterns.

This chapter will focus on various communication patterns that are prevalent in microservices architecture. We will cover two primary approaches used for communication such as request/response and pub/sub communication and then explore various pub/sub communication protocols, such as RabbitMQ and Kafka. We will set up Kafka with Azure Event Hubs and show how to integrate services with it. Understanding this will lead you to learning new ways of building applications that you can take for your day-to-day work directly.

Approaches to communication

In a monolithic application, we can generally have components communicate with others using embedded code that initiates an action to reach out and expect a response. There are methodologies that exist within monolithic scenarios that work in this fashion, but there might be issues when it comes to scale. Let's posit a traditional monolithic scenario where you leverage embedded code within services to directly call and communicate with other services. When stringing these calls together, there isn't an effective handler that can sustain the line of communication and with greater numbers and scale of these calls, the more inefficient the communication system becomes. Without structure to perform an end-to-end scenario, one inadvertently causes more overhead than needed. Let's assuage this and introduce some communication basics to understand how we address communication patterns within microservices.

Synchronous versus asynchronous communication

In the microservices architecture, we generally think of two particular kinds of communication—synchronous and asynchronous. Synchronous communication is based on real-time awareness, meaning that two separate components should be able to use the mechanism or protocol in place to instantly send and receive messages from the other side. HTTP is a very common example of a synchronous communication protocol.

Asynchronous communication is more focused on sending messages without an immediate response expected. Unlike the synchronous communication protocol that establishes an actual connection with a component and expects a direct response, asynchronous methods just send the message through a broker without having to wait for a response. This encourages one-to-many and many-to-one connections that aren't resigned to the rigidity in synchronous communication. An example of an asynchronous communication protocol is AMQP (Advanced Messaging Queuing Protocol), which is shown in Figure 6-1.

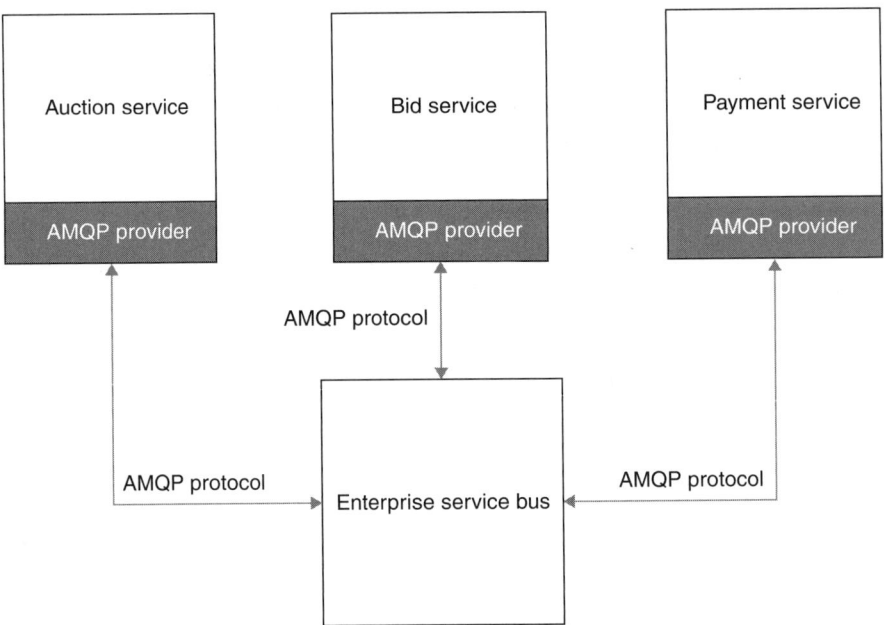

FIGURE 6-1   AMQP communication in the context of the OAS back-end microservices

Figure 6-1 shows an example of AMQP communication in the context of the Online Auction System (OAS) back-end microservices, though note that we didn't use AMQP in our actual implementation.

> **NOTE**   Diversity in formats
>
> With the diversity in communication protocols, there is also diversity in the content that is passed within these communication funnels. You can use file formats such as JSON and XML, but there are binary options and non-standard options as well that allow you to choose specific formats for particular use cases. For the OAS, we utilized JSON as our message format.

For a complete breakdown of synchronous and asynchronous communication attributes, see Table 6-1.

TABLE 6-1 Synchronous versus asynchronous communication

Synchronous Communication	Asynchronous Communication
Real-time awareness	Delayed communication
Single message with an instantaneous answer if the service is available	Send messages with a broker without wait times
Protocols such as HTTP or HTTPS	Protocols such as AMQP
Mechanisms such as request/response	Mechanisms such as pub/sub

Request/response communication

Let's dive deeper into one form of synchronous communication, request/response communication. Before diving deeper into the protocol itself, let's also understand how to evaluate its complexity. The simple way to understand this is that the more components that are involved in the communication, the more complex the protocol is. Furthermore, you also want to tailor protocols based on use case. This means that for simpler scenarios where we have a single component communicating with another component, synchronous communication would be a good fit to send out a request and expect a response. In fact, this is the basis of the communication protocol called the request/response which is a form of synchronous communication that is best suited to protocols such as HTTP and HTTPS, along with a web API such as the REST API.

Figure 6-2 shows an example of the request/response mechanism for the Online Auction System. The bid service is trying to communicate with the auction service, perhaps updating the latter on the most recent bid values that are associated with a particular active auction. This direct communication function with an HTTPS call is funneled through Azure API Management (APIM), which is a PaaS service in Microsoft Azure that is focused on application endpoint management with the use of tools such as API Gateways. This request is processed and sent over Azure APIM to transact directly with the auction service, which will send back a response confirming the completion of the request. We will go into detail about API gateways and Azure APIM in Chapter 8.

FIGURE 6-2 Communication between the bid and auction services

Although this is not the method we used in the OAS application, we made an important design decision to leverage Azure APIM to use tools such as the API Gateway, which is important for communicating with several of these microservices. This is considerably easier to implement for your environment, though it might not be the best fit for use in conjunction with microservices.

Pub/sub communication

In more complex scenarios, we start thinking about a single component having to correspond with multiple components. To contextualize it with microservices, a single microservice communicates with several other partner microservices. Here, the best fit would be asynchronous communication, which can send out messages to different receivers through a singular message broker.

The publish/subscribe method, frequently called "pub/sub," is a commonly used solution in these types of scenarios. As we discussed earlier in this chapter, it would not make sense to have embedded calls to communicate with our services, and the request/response method might not be the best fit either. While the request/response method can better perform synchronous communication, scaling problems are introduced when several services need to speak to one another.

The pub/sub method is particularly helpful here because it works very well in conjunction with an event-based architecture that uses messaging brokers and strings together multiple communications for an end-to-end scenario.

Figure 6-3 shows how pub/sub communication works with the OAS. The auction service, with its associated database, publishes a particular event over an event bus. This is our pub/sub channel—the message broker—which helps shuttle these communications to the necessary services.

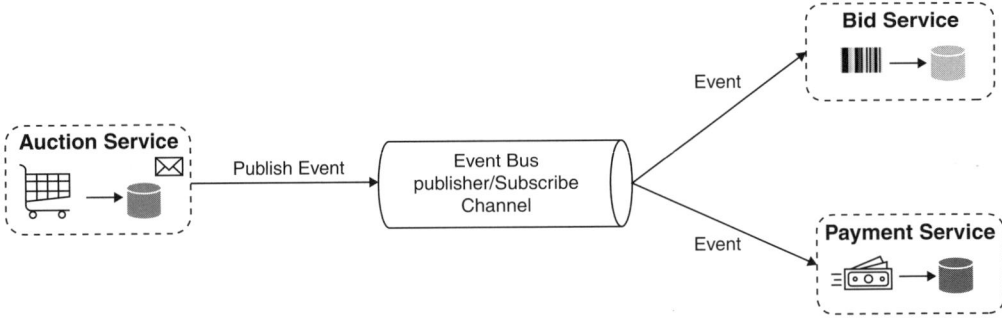

FIGURE 6-3 Asynchronous communication using pub/sub communication between the auction service and the bid and payment services

The event is pushed out asynchronously to the bid and payment services, which store the event in their databases to process and manipulate the data sent from the auction service.

The processing aspect works like a chain of reactions. For example, the event processed in the auction service can result in a new bid service action. A pattern emerges in which this mechanism is repeatedly kicked off to string together several events and create a computation where the user sees a tangible result in the application. The nature of asynchronous communication enables this end-to-end communication, whereas synchronous communication may might be the best choice when communicating between several microservices. This resulted in us leveraging pub/sub communication for the OAS.

The best communication approach for microservices

As mentioned earlier, it might be difficult to create a cohesive communication flow using the request/response methodology. This is particularly true because of its transactional way of operating in which a prompt and a response are needed for a complete line of communication. Chaining these one-to-one events can slow the overall process and raise potential time-out errors. We can see why this happens in Figure 6-4.

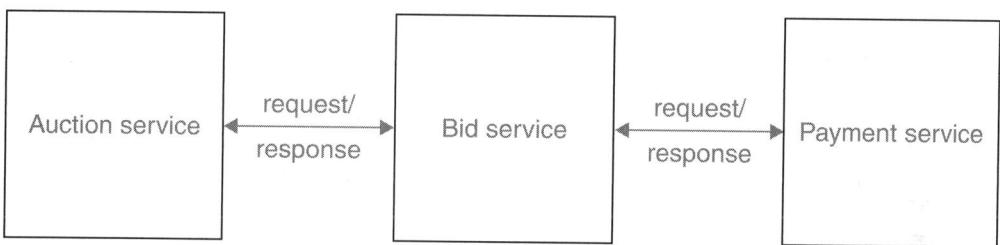

FIGURE 6-4   The fallacy in using combined request/response flows

Each segmentation within this example would result in timeout errors as a result of waiting for the response from the next microservice. This is why it's not a best practice and why we haven't implemented it for the OAS. Asynchronous communication practices like pub/sub are much preferred in microservices. By not having a transactional communication protocol—especially one that doesn't have to wait for a response—we can create a cohesive chain for an end-to-end scenario using well-established patterns.

Because we are publishing events, structuring each service with its own data store allows one to store information to be acted on and then push it over to the next service and process it until complete.

Figure 6-5 shows the **Create Auction** tab in the Online Auction system. After filling out the required fields, other users can see this within the **Active Auctions** tab.

From the standpoint of our OAS, a generic user flow can be that User 1 initiates an auction through our app's front-end. This act of filling out the form fields in the **Create Auction** workflow kicks off the respective microservice in the back-end. As you can see in Figure 6-5 below, User 1 has filled out the relevant fields to initiate an auction to sell a car for $1,000.

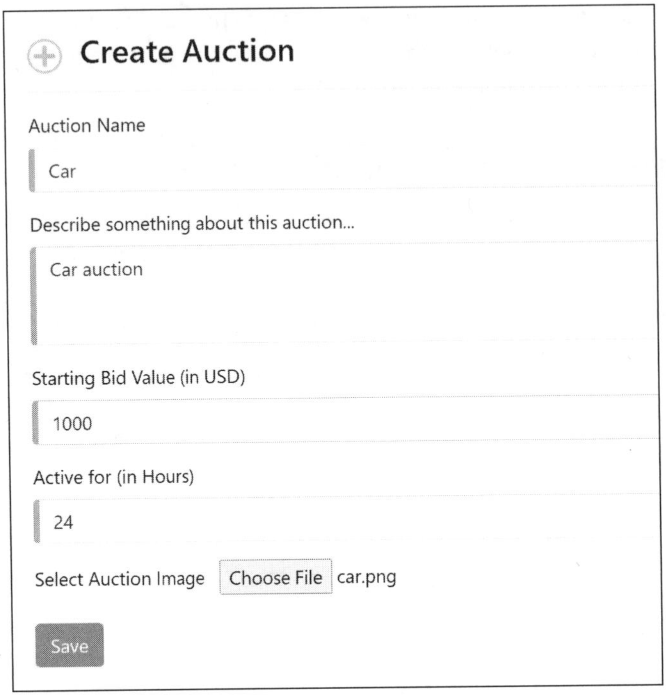

Create Auction

Auction Name

Car

Describe something about this auction...

Car auction

Starting Bid Value (in USD)

1000

Active for (in Hours)

24

Select Auction Image   Choose File   car.png

Save

FIGURE 6-5   The Create Auction workflow in the Online Auction System

Now, to continue the user flow, User 2 sees the auction initiated by User 1 in the **Active Auctions** section and decides to bid on it. In Figure 6-6, we see that the bid can be submitted and is part of the **Active Auctions** tab. Ultimately, if this bid stands as the highest bid, the payment service will then kick in to confirm the winning bid and finalize the results of the auction. Here, we have a specific workflow with an auction being posted and then generating an entry within the **Active Auctions** tab for a delineated amount of time.

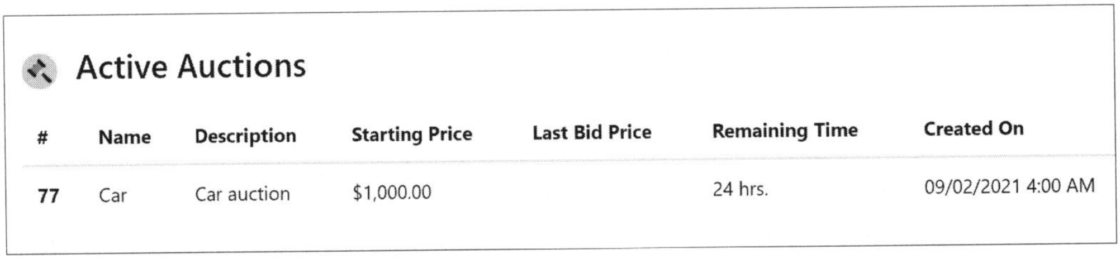

Active Auctions

#	Name	Description	Starting Price	Last Bid Price	Remaining Time	Created On
77	Car	Car auction	$1,000.00		24 hrs.	09/02/2021 4:00 AM

FIGURE 6-6   Active Auctions workflow within the Online Auction system

The table shown in Figure 6-6 is filled in with the values of the Auction Service DB schema, and other users can use this information to submit bids of certain value.

Seeing this entry on the **Active Auctions** screen, User 2 can choose to bid on the auction, which is shown in the in the **Make A Bid** workflow flow in Figure 6-7.

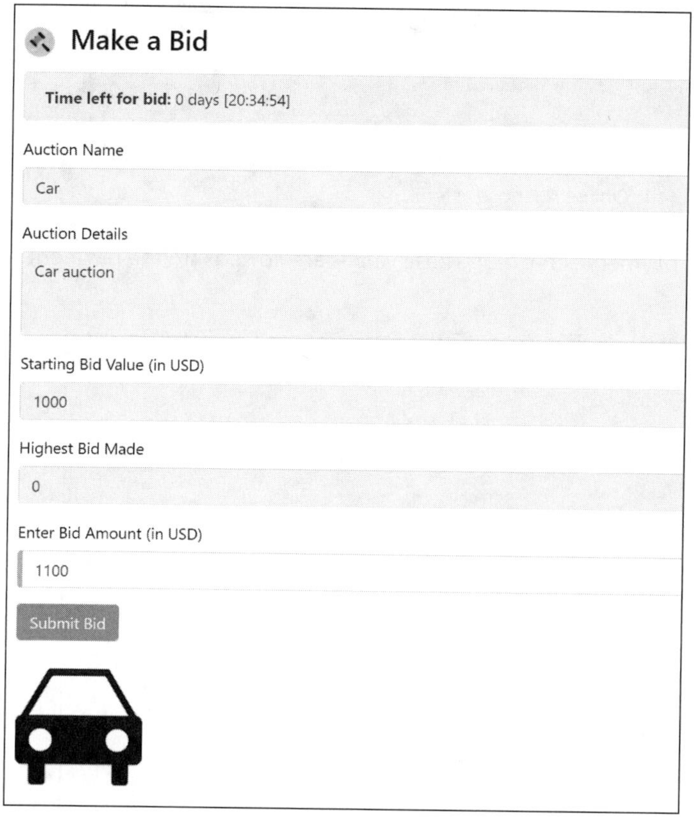

FIGURE 6-7 Make A Bid workflow

Figure 6-8 shows the **My Auctions** tab in the OAS. This table represents the view from **User 1's** perspective who has initiated an auction. The figure shows the bid entered by **User 2** on this auction from Figure 6-7. This is the result of multiple services communicating with each other to create a cohesive auction bidding experience.

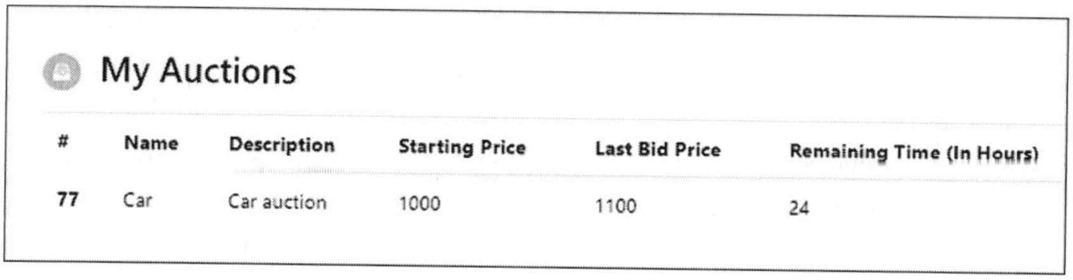

#	Name	Description	Starting Price	Last Bid Price	Remaining Time (In Hours)
77	Car	Car auction	1000	1100	24

FIGURE 6-8 The My Auctions tab in the OAS

Figure 6-9 shows that the Winning Bids table is filled in with the values of the Auction Service DB schema. This event would occur after the auction expires and the remaining highest bid is chosen as the winner. The information of that highest bid is then posted here with the option for the winner—in this case, User 2—to make the payment and complete the auction workflow.

Winning Bids

#	Name	Description	Starting Price	Winning Price	Payment Status
77	Car	Car auction	1000	1100	Make Payment

FIGURE 6-9 Winning Bids section of the Online Auction system

Figure 6-10 shows the OAS's payment service UI. These values are flowed into the payment service database schema.

Auction Payment

Auction Name

Car

Auction Details

Car auction

Payment Details

CARD NUMBER

Valid Card Number

EXPIRATION DATE

0   .   0

CV CODE

CV

Final Payment: $1100

Pay

FIGURE 6-10 The payment service UI in the OAS

Although we elaborated the interaction of the different services with a small set of users with a simplistic example, attempting to increase the scale of this system with more users, transactions, and communication will introduce complexity. Thankfully, we have used some of the methodologies mentioned in this chapter to propagate end-to-end communication. Each action by a user triggers events that pass to each of the various services, resulting in a cohesive workflow. You will also notice that we have used the SAGA and CQRS patterns defined in Chapter 3, "Build microservices architecture and use Azure." In fact, our implementation contains a shortcut to reduce the overhead in communication by having the highest bid values stored in the auction service database schema. Essentially, this would mean particular components of the bid service database schema would be in the auction service. By doing this, the final transaction of this OAS workflow will directly pull the information for a winning bid straight from the auction service, rather than having to coordinate between the auction, bid, and payment services.

Ultimately, it is important to architect out all of your components first before deciding the nuanced approach of which modes of communication to use. What we used with OAS is a great way to approach it, but it's not the only way, and hopefully, we have illuminated the optionality you have in this area. Utilizing these patterns and best practices along with these types of communication helps us achieve the best functionality.

Pub/sub communication technologies

Now that we have learned that pub/sub communication is one of the best options to use from the standpoint of microservices, let's do a deep dive into some of the technologies that will help us realize this in an actual application.

Apache Kafka

Apache Kafka is an open-source project that is focused on message brokering and communication. The project is widely known and thus has a massive following that has generated many libraries, APIs, and connectors that enable its use into applications for communicating between components of large and complex on-premises, cloud-native, and hybrid architectures.

Kafka is particularly relevant to our conversation because it is a technology that enables the pub/sub communication model in practice.

Construct of Kafka

Kafka is a cluster-based service made up of several virtual machines. You can run the service as needed in any location (cloud or on-prem) and can leverage it as a message broker to essentially reroute and "publish" source messages to other components that will subsequently consume these events. Ultimately, there are producers and consumers that work to publish and subscribe to events to complete a communication flow. This is why it is so reflective of the pub/sub communication protocol. The breakdown of a Kafka cluster in its very base components is shown in Figure 6-11.

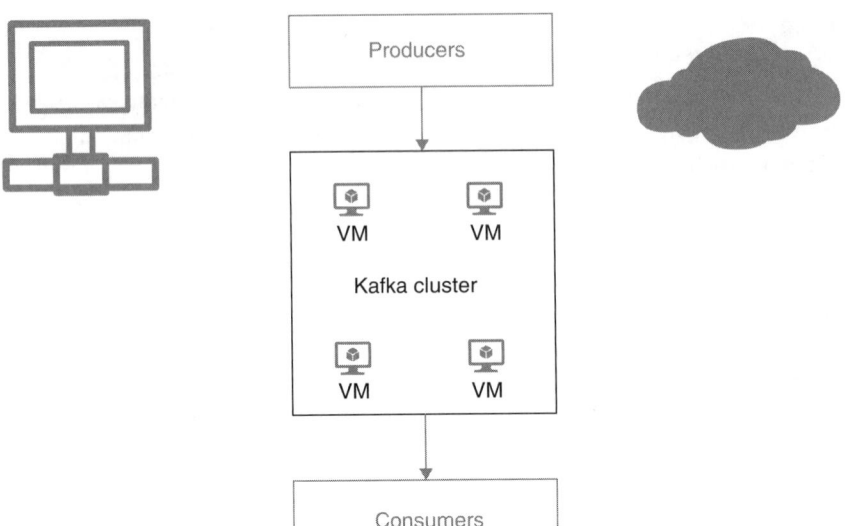

FIGURE 6-11 Kafka cluster

When considering the standalone open-source version of Kafka and its amazing flexibility, we also have to consider some of the potential design downsides. In particular, the management of these Kafka clusters can become difficult.

Management of Kafka clusters and brokers introduces overhead that can become quite laborious. The sheer number of options you have with this management can increase the complexity of your messaging customizations within your architecture and introduces decision fatigue. Aside from the inundation of options, the flexibility to create intricate messaging channels means that there are lots of customizations and configurations needed for a particular app. As your application scales and grows, you will need to change these configurations yet again. Thus, in many IT environments, we see that there is usually a dedicated person or team responsible for maintaining this infrastructure.

Furthermore, this diminishes a primary benefit of using microservices, which is to be modular and to push code quickly. However, you now have to reconcile the code changes done with Kafka meticulously, and thus, you've lost the hard-earned agility and flexibility of microservices. Interweaving this with the logical code you have becomes difficult, which puts us back at square one because we don't have an architecture to meet these best practices.

The maintenance of the Kafka clusters and infrastructure is a major pain point, though the fundamental Kafka protocol itself is still a very valuable tool to leverage. You might wonder if you could use the Kafka protocol through a managed service. Wouldn't this address the issue of overhead and decision fatigue? You would be correct in that assessment, and this methodology is actually the solution that we have opted for using with the OAS. We did this by leveraging the Azure Event Hubs with Kafka protocol.

Azure Event Hubs

Before we go into detail on the particular integration we used within our application, we want to first go over Azure Event Hubs as a whole. We touched on this briefly before, but the use of Azure Event Hubs as our enterprise bus was key to the success of the OAS. While there are many options you can use for an enterprise, we chose to use Azure Event Hubs in the OAS.

Event Hubs acts as an event-ingestion service that takes in information from different sources that can then be processed for a cohesive operational flow. In particular, this is a cloud-based feature that allows us to collect information from our OAS application that is, of course, cloud-native and provides the benefits of an extant integration. Event Hubs cover scenarios and concepts using the following features:

- Fraud detection
- Logging
- Data pipelines
- Play-by-play dashboards
- Data storage
- Transactions
- Telemetry

> **NEED MORE INFO?** Event Hubs
>
> For information on Event Hubs, see the Microsoft Documentation at
> *https://docs.microsoft.com/en-us/azure/event-hubs/event-hubs-about*.

Utlimately, Azure Event Hubs is a fantastic Platform-as-a-Service (PaaS) tool to help ingest and process events, and this particular nature helps make it a great tool when used with Kafka.

Event Hubs with Apache Kafka

We ended up leveraging Event Hubs with Apache Kafka to get the best of both worlds. Instead of worrying about the complications in leveraging our own Kafka cluster, we used the Event Hubs feature to help account for these problems. Essentially, this feature allows a Kafka endpoint to stand-in for the Kafka cluster. This was a great way to leverage a serverless tool to absolve us of the tasks needed while maintaining a Kafka cluster, providing simplicity to our communication methodology

From an operational standpoint, this was a great decision because Event Hubs and Apache Kafka are quite similar from the theoretical and component standpoints. For example, the concepts of the Kafka cluster as well as a service bus topic are similar to the concepts of the Event Hub namespace and the Event Hub, respectively. This means that the architecture using the Event Hub namespace uses similar components to the architecture using the Kafka cluster, easily enabling communication use cases.

Now if these tools and architectures are similar, why did we end up using this feature set over a Kafka cluster? The biggest benefits we can see from this feature compared to Apache Kafka is the serverless nature and scale. As we mentioned before, the cluster management of

Kafka is different to contend with, and a PaaS tool or managed service such as Event Hubs truly removes that overhead, especially within a DevOps and an agile team setting. Aside from the infrastructure management perspective, the use of Azure enables scale to handle more event throughput and consumption, and this is definitely a pain point to consider, especially if Kafka was deployed on-premises where there are additional measures needed to scale.

Because we were building the OAS application on Azure, we thought it was one of best ways to leverage the power of the cloud in conjunction with our design choices on communication. As a result, it worked very well for us, though it isn't the only consideration for pub/sub communication technologies that you can use for your applications.

RabbitMQ

Your environment will likely be different from ours in terms of building an application, and so other tools, such as RabbitMQ, should be considered in addition to Kafka and Event Hubs.

RabbitMQ is similar to Kafka in that it also is an open-source message broker and communication technology. RabbitMQ can utilize the AMQP protocol with several other protocols that are suited for pub/sub communication. RabbitMQ is similar to Kafka in that you can cluster RabbitMQ, focus on reliability and availability, and use many APIs, connectors, integrations, and so on. In fact, it has a large community following, too, as seen from how feature-rich it is. RabbitMQ is a tool that is the longstanding go-to tool for application communication and that is frequently used by architects, developers, consultants and more.

RabbitMQ can use many protocols and was originally used as a communication tool with antiquated monolithic, and service-oriented architectures. However, it has adapted and provides functionality similar to what we see in Kafka.

In many senses, RabbitMQ is a better choice than Kafka. RabbitMQ provides the perfect balance of flexibility thanks to a robust set of protocols, including synchronous communication. RabbitMQ also addresses Kafka's biggest weakness because it doesn't need nearly as much compute-intensive resources or physical infrastructure to run.

In addition, there are nuanced differences when using RabbitMQ that are prevalent within the mechanisms used in the technology along with niche use cases as well. That being said, it is still best to use RabbitMQ on legacy applications. Older pub/sub protocols and modern applications with streaming use cases should leverage Kafka. An argument could still be made that RabbitMQ can also address streaming use cases as well, albeit not as well because RabbitMQ is an older technology that has been retrofitted with integrations to other technologies to work for the same use cases, whereas Kafka was designed with event-streaming in mind.

Thus, we decided to utilize Kafka rather than RabbitMQ. Although both are great open-source representatives, the integration with Azure Event Hubs and Kafka provided greater simplicity to us and thus was one of the deciding factors. Be sure to be comprehensive when deciding which tool to use with your applications. Be sure to leverage the benefits available to you from your cloud provider, as we had done with Azure.

This being said, we still want you to understand how this is deployed within a microservices architecture and so we will demonstrate this by walking you through how we set up our communication patterns in the OAS.

Set up Kafka to establish pub/sub communication

In this section, we will explore the steps to set up Kafka using Azure Event Hubs and produce messages from a Java Spring Boot application, while a .NET Core application will be used as a consumer. Essentially, we will be modifying the bid service that we discussed in Chapter 5 with the necessary code to produce messages in Kafka. We will also go through the steps regarding the .NET Core Hosted Service, which is the consumer for our messages, and some information on how to set this all up to be resilient.

Infrastructure setup

Before we dive into any of the configuration and code topics, we want to address how to set up some of the components that we used in our application. There are various options available in Azure Marketplace to set up Kafka. For example, a Kafka cluster is available from Bitnami, Azure HDInsight, and so on, though as we mentioned earlier, we went with Azure Event Hubs for Apache Kafka instead.

Within Azure Event Hubs, we are leveraging the Kafka endpoint, which means you just need to provide the new configuration values, and we are good to go. To start setting this up, navigate to the **Create A Resource** section within the Azure portal's side bar and type **Event Hubs** to provision this resource. See Figure 6-12.

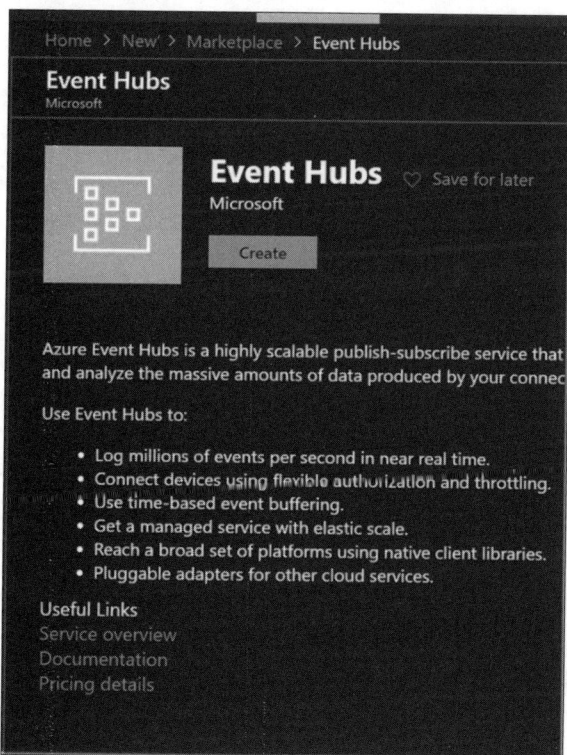

FIGURE 6-12   Azure Event Hubs provisioning

Click **Create** to open the **Create Namespace** window and fill in the following fields for **Subscription, Resource group, Namespace name, Location, Pricing tier,** and **Throughput Units,** as shown in Figure 6-13. In order to enable Kafka, you must use a Standard pricing tier or higher.

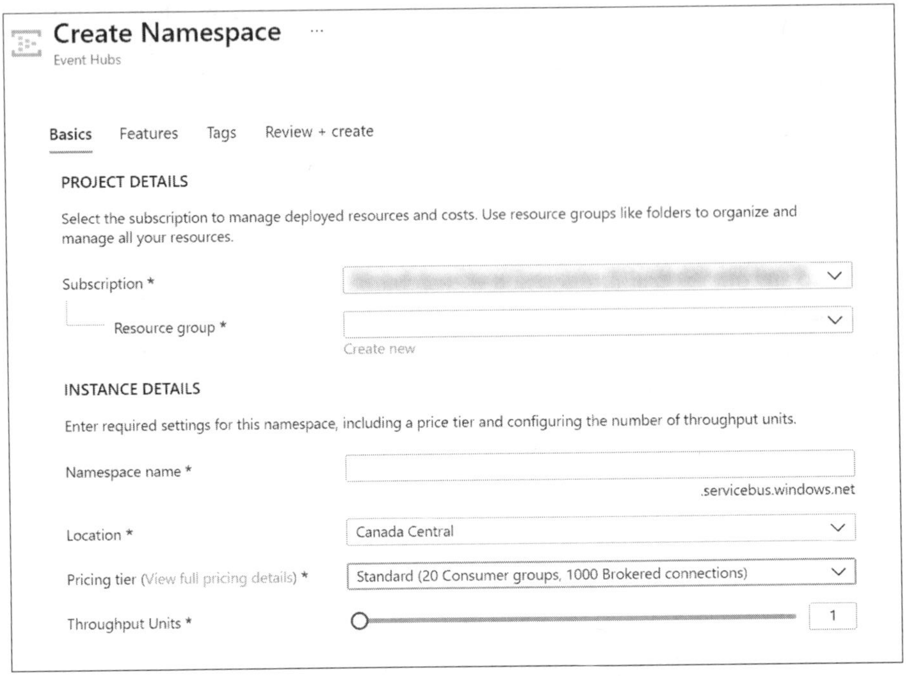

FIGURE 6-13 Create Namespace

Once the Kafka namespace is created, topics can be created by selecting the **Event Hubs** option under **Entities** and clicking the **+EventHub** option. From here, you can go to the **Create Event Hub** section, as seen in Figure 6-14, and fill in the event hub's **Name, Partition Count,** and other options to provision this resource.

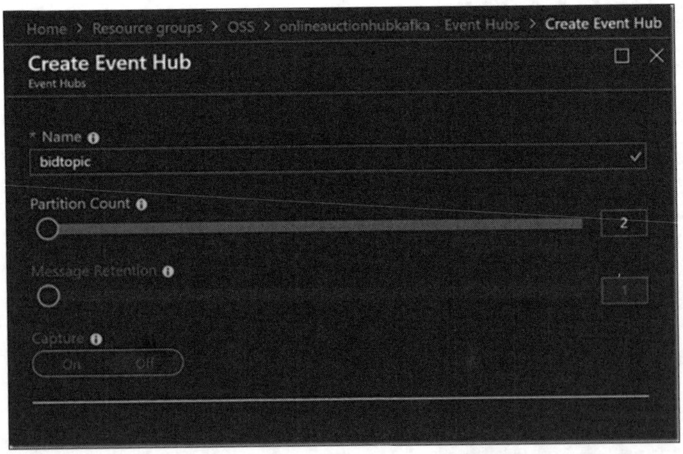

FIGURE 6-14 Create Event Hub

Once the topic is created, we can start producing messages for that topic and consume them, which is all we need to do from the provisioning side! Before we go into more detail into the code for producing messages in Kafka, we want to show an overall view of this part of the architecture, so you understand why we are performing these particular actions. Figure 6-15 shows an architectural view of the communication method we are setting up between the bid and auction services. We are, of course, leveraging Azure Event Hubs, but we also have the listener service that is built from a WebJob that we will discuss later in this chapter.

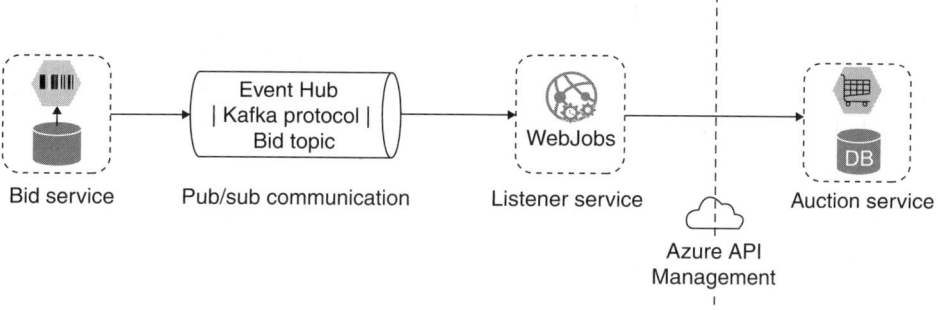

FIGURE 6-15 Architectural view of the communication method between the bid and auction services

In the example shown in Figure 6-15, we see that the bid topic was set up in Kafka to store bid messages. Then the Kafka client was used in Java to submit messages in JSON format to the bid topic. We also ended up using the Confluent library in .NET Core to read Kafka messages from the bid topic and invoke the Azure APIM endpoint to communicate with the auction service.

In the next section, we will go into detail about setting up the producer code.

Setting up the producer: Adding Kafka support in the Java application

We already created the bid service in Chapter 5. In this chapter, we will modify and add code to produce messages. First, we will add the dependencies required to use Kafka in the Java application, such as a web API that is built on the Java Spring Boot framework that exposes an endpoint. Once the user hits that particular endpoint, we want it to read a certain value and push it to the Kafka topic. To add Kafka support, edit the pom.xml file to add the following dependency:

```
<dependency>
<groupId>org.apache.kafka</groupId>
<artifactId>kafka-clients</artifactId>
<version>0.11.0.0</version>
</dependency>
```

Next, create the `producer.config` file and add the connection string, Kafka server endpoint, and so on. Here is the configuration for the `producer.config` file, which was added to the /src/main/resources folder.

```
bootstrap.servers= {serverendpoint}:9093

security.protocol=SASL_SSL

sasl.mechanism=PLAIN
sasl.jaas.config=org.apache.kafka.common.security.plain.PlainLoginModule required
username="$ConnectionString" password="{password}";
```

To obtain the {serverendpoint} and {password} values, go to Event Hub and click the **Shared Access Policies** tab. Choose the policy and copy the connection string-primary key value. This whole value is the password. You can then extract the server endpoint from the same value and provide it for the `bootstrap.servers` key:

```
{youreventhubnamespace}.servicebus.windows.net
```

Next, add the code snippet shown in Listing 6-1 to send messages to Kafka and use the configuration values from the `producer.config` file:

LISTING 6-1 Modifying the CreateBid method of BidController to publish messages using Kafka protocol

```java
try
  {
            /*
            * Defining producer properties.
            */
            Properties properties = new Properties();
            telemetryClient.trackTrace("Sending messages to Kafka bid topic");
            properties.load(new FileReader("src/main/resources/producer.config"));
            properties.put(ProducerConfig.KEY_SERIALIZER_CLASS_CONFIG, LongSerializer.
            class.getName());
            properties.put(ProducerConfig.VALUE_SERIALIZER_CLASS_CONFIG, StringSerializer.
            class.getName());
            KafkaProducer<Long, String> producer = new KafkaProducer<>(properties);
            long time = System.currentTimeMillis();
            Gson gson=new Gson();
            String bidObjectJson  = gson.toJson(doc);
            final ProducerRecord<Long, String> record = new ProducerRecord<Long, String>
            ("bidtopic",time, bidObjectJson);
            producer.send(record, new Callback() {
                public void onCompletion(RecordMetadata metadata, Exception exception) {
                    if (exception != null) {
                        System.out.println(exception);
                        System.exit(1);
                    }
                }
            });
```

```
            telemetryClient.trackTrace("Message sent to Kafka bid topic");
        }
        catch (Exception ex)
        {
            System.out.print(ex.getMessage());
            throw ex;
        }
```

The above code is included in the BidController code in Chapter 5, "Microservices on containers." Initialize the KafkaProducer object by passing the producer.config properties, send a message using the producer.send method, and pass a ProducerRecord object.

Setting up the Consumer: Develop Kafka Listener service with .NET Core Hosted Service

The listener service is the background service hosted as a WebJob to listen for Kafka events. The primary goal of this service is to provide integration between microservices. This service captures events that originate from Kafka and call the appropriate service API to transfer data.

To visualize this, think about when a bid is made by the user. The bid service creates an entry in a bid topic and provides bid details. This hosted listener service will then capture that event, read the message from the bid topic, and call the auction service API with the data received. The auction service then extracts the values such as bidUserId and bidPrice and updates the auction table. This way whenever the bid is made the auction table gets updated with the last bid offer, including the user who made that bid.

Now in the technical implementation, for background tasks within .NET specifically it is best to use a hosted service to run as a background service. This hosted service is actually an interface that we must implement in order to process the messages that are coming from bid service, as previously shown in Figure 6-15. This will function like a message queue which is reflective of some of the design components we have mentioned earlier in this chapter. In essence, we extend this interface to create a hosted service that can run In conjunction with the consumer code. We will walk through this next.

To start, create a new .NET Core project to build the Kafka listener service. We will not be covering the steps to create a .NET Core project here in this book, but do use the public documentation available to get a project kickstarted in your respective IDE or editor. Once the project is created, we will add a NuGet package to consume messages from Kafka. To do this, there are many Kafka libraries available. In our case, we used the Confluent Kafka library.

To add the Confluent Kafka library, open the NuGet package manager in Visual Studio and search for the **Confluent Kafka** library, as shown in Figure 6-16.

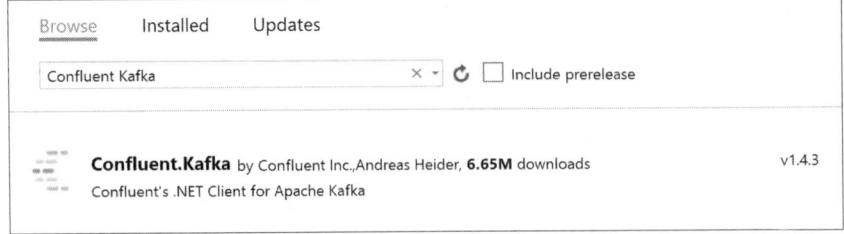

FIGURE 6-16 Confluent Kafka library

Before we start building out the hosted service, we will create a KafkaConsumer class. Listing 6-2 shows the complete code snippet of KafkaConsumer class used to consume messages from Azure Event Hubs using the Kafka protocol.

LISTING 6-2 KafkaConsumer Class

```
using Confluent.Kafka;
using Microsoft.Extensions.Configuration;
using Newtonsoft.Json.Linq;
using System;
using System.Collections.Generic;
using System.Linq;
using System.Threading;
using System.Threading.Tasks;
namespace EventListener
{
    public class KafkaConsumer
    {
        private KafkaHttpClient _client;
        private IConfigurationRoot _configuration;
        public KafkaConsumer(KafkaHttpClient client, IConfigurationRoot configuration)
        {
            _client = client;
            _configuration = configuration;
        }
        public void ConsumeMessages(string topic)
        {
            var config = new ConsumerConfig
            {
                BootstrapServers = _configuration["BootstrapServers"],
                SecurityProtocol = SecurityProtocol.SaslSsl,
                SocketTimeoutMs = 60000,
                SessionTimeoutMs = 30000,
                SaslMechanism = SaslMechanism.Plain,
                SaslUsername = "$ConnectionString",
                SaslPassword = _configuration["SaslPassword"],
              GroupId = _configuration["GroupId"],
                EnableSslCertificateVerification=false,
```

```
            AutoOffsetReset = AutoOffsetReset.Earliest,
            BrokerVersionFallback = "1.0.0",
        };
        using (var consumer = new ConsumerBuilder<Ignore, string>(config).Build())
        {
            Logger.Instance.LogMessage("Subscribing topic");
            consumer.Subscribe(topic);
            CancellationTokenSource cts = new CancellationTokenSource();
            Console.CancelKeyPress += (_, e) => {
                e.Cancel = true; // prevent the process from terminating.
                cts.Cancel();
            };
            try
            {
                Logger.Instance.LogMessage("Starting loop to check for Kafka messages");
                while (true)
                {
                    Logger.Instance.LogMessage("Inside While loop to
                    check for Kafka messages");
                    try
                    {
                        var cr = consumer.Consume(cts.Token);
                        Logger.Instance.LogMessage($"Got one message value is
                        {cr.Value}");
                        dynamic data = JObject.Parse(cr.Value);
                        string bidId = data.bidid;
                        string customerId = data.userId;
                        decimal bidAmount = Convert.ToDecimal(data.bidAmount);
                        int auctionId = Convert.ToInt32(data.auctionId);
                        Logger.Instance.LogMessage($"Calling Update Auction
                        for Bid for {cr.Value}");
                    }
                    catch (ConsumeException e)
                    {
                        Logger.Instance.LogMessage($"Error occured: {e.Message}
                        Stack  Trace: {e.StackTrace}");
                    }
                }
            }
            catch (OperationCanceledException)
            {
                consumer.Close();
            }
        }
    }
}
```

In Listing 6-2, we added a `ConsumeMessages` method to receive messages from the topic. We have some of our configuration pieces here that are being filled in with information from our `appsettings.json` file. After setting this up, we are subscribing the topic and then use a `while` loop to continuously check for Kafka messages. When a message arrives, we are parsing out the relevant values from that message. Before we get to the final step, we want to have the other pieces in place to demonstrate how this works end to end. Let's now proceed with the hosted service.

BidHostedService class

Now that we have the mechanism in place to deal with messages, let's now focus on listening for them and actually using the consumer. We do this by implementing the `IHostedService` interface for .NET Core. This interface has two key methods we are concerned about: `StartAsync` and `StopAsync`. As mentioned in the Microsoft documentation, `StartAsync` is "Triggered when the application host is ready to start the service," and `StopAsync` is "Triggered when the application host is performing a graceful shutdown." Listing 6-3 shows our `BidHostedService.cs` code for the OAS:

LISTING 6-3 BidHostedService class

```
public class BidHostedService : IHostedService
    {
        private KafkaConsumer _consumer;
        private IConfigurationRoot _configuration;
        public BidHostedService(KafkaConsumer consumer, IConfigurationRoot configuration)
        {
            _consumer = consumer;
            _configuration = configuration;
        }
        public Task StartAsync(CancellationToken cancellationToken)
        {
            Logger.Instance.LogMessage("Starting Listener for BidTopic");
            _consumer.ConsumeMessages(_configuration["BidTopicName"]);
            return Task.CompletedTask;
        }
        public Task StopAsync(CancellationToken cancellationToken)
        {
            Logger.Instance.LogMessage("Stopped Listener for BidTopic");
            return Task.CompletedTask;
        }
    }
```

As you see here, we started by creating the attributes for the class and the constructor, which connects to the information created within the consumer class. Afterward, we implement the two mandatory functions as per the interface. You can now add this hosted service and related dependencies within your main program file, which we created as a `Program.cs` class. We demonstrate this in Listing 6-4.

LISTING 6-4 Main Program.cs file

```
static async Task Main(string[] args)
        {
            // Build configuration
            var configuration = new ConfigurationBuilder()
                .SetBasePath(Directory.GetParent(AppContext.BaseDirectory).FullName)
                .AddJsonFile("appsettings.json", false)
                .Build();
            await new HostBuilder()
            .ConfigureServices((hostContext, services) =>
            {
                services.AddSingleton<IConfigurationRoot>(configuration);
                services.AddSingleton<KafkaConsumer>();
                services.BuildServiceProvider();
                services.AddHostedService<BidHostedService>();
            })
            .RunConsoleAsync();
            Console.ReadLine();
        }
```

So far, we have developed the hosted service that calls the KafkaConsumer to consume messages, but this is not our only goal. After reading the message with these services, we still want to call the auction service API. In this particular case, we can use something like an HttpClient object to call it. However, before we proceed to that step, we should first explore resiliency patterns in order to create a custom HttpClient class. The process in going through this will help us better understand resiliency concepts, another fundamental topic in thinking about microservices communication.

Resiliency: Retry and circuit breaker patterns

The primary goal of a listener service is to provide integration between services, which is done mostly over an HTTP/HTTPS protocol. Once the message is published to the Kafka topic, this service picks up that message and calls the corresponding API for integration.

> **NOTE** Resiliency
>
> Before we get into the protocol, we need to take a quick aside on resiliency. Resiliency is an important factor when working with this kind of service. Resiliency is the ability of a system to gracefully handle and recover from failures. It is one of the core patterns to be used with microservices architecture where an application is decomposed into multiple services, and each service communicates with other services over HTTP/HTTPS endpoints. This is especially important when coding it in conjunction with this listener service because it is a vital integration that is constantly listening to the flow of event. Being unable to adequately handle exceptions and time-outs will result in a many issues and can create a potentially awful user experience.

To accommodate resiliency in the listener service, we will use the retry pattern in conjunction with the circuit breaker pattern.

RETRY PATTERN

The retry pattern is used where we need to retry the faulted service several times in order to get a valid response and fails after the maximum limit of retries. If an application detects failure, it will retry the service endpoint using following strategies:

- **Cancel**   If the fault is not a transient fault, we should cancel rather hitting the service endpoint repeatedly.
- **Retry**   If the fault is a transient fault, we should call it immediately to get the valid response.
- **Retry after delay**   If the fault is a transient fault, we can also call the service endpoint after some delay or time interval. This is mostly used when there is some network failure.

The application should wrap all attempts to access a remote service in code that implements a retry policy matching one of the strategies listed above. Requests sent to different services can be subject to different policies. Some vendors provide libraries that implement retry policies, where the application can specify the maximum number of retries, the time between retry attempts, and other parameters.

An application should log the details of faults and failing operations. This information is useful to operators. If a service is frequently unavailable or busy, it's often because the service has exhausted its resources. You can reduce the frequency of these faults by scaling out the service. For example, if a database service is continually overloaded, it might be beneficial to partition the database and spread the load across multiple servers.

CIRCUIT BREAKER PATTERN

The circuit breaker pattern is used to detect failures and encapsulate the logic of preventing constantly recurring failures, maintenance, or temporary external system or unexpected system failures.

In a distributed environment, calls to remote resources and services can fail because of transient faults, such as slow network connections, timeouts, resources being overcommitted or temporarily unavailable. These faults typically correct themselves after a short period of time, and a robust cloud application should be prepared to handle them by using a strategy such as the retry pattern.

However, there can also be situations where faults are caused by unanticipated events and that might take much longer to fix. These faults can range in severity from a partial loss of connectivity to the complete failure of a service. In these situations, it might be pointless for an application to continually retry an operation that is unlikely to succeed, and instead, the application should quickly accept that the operation has failed and handle this failure accordingly.

A circuit breaker acts as a proxy for operations that might fail. The proxy should monitor the number of recent failures that have occurred and use this information to decide whether to allow the operation to proceed, or simply return an exception immediately.

The proxy can be implemented as a state machine with the following states that mimic the functionality of an electrical circuit breaker:

- **Closed** The request from the application is routed to the target service. If it's used in conjunction with a retry pattern, then it maintains a count of the number of recent failures, and if the call to the target service is unsuccessful, retry pattern increments the retry counter. If the number of recent failures exceeds a specified threshold within a given time period, the proxy is placed in an open state. At this point, the proxy starts a timeout timer. When this timer expires, the proxy is placed into the half-open state.

- **Open** The request from the application fails immediately and an exception is returned to the application.

- **Half-Open** A limited number of requests from the application are allowed to pass through and invoke the operation. If these requests are successful, it's assumed that the fault that was previously causing the failure has been fixed and the circuit breaker switches to the closed state. (The failure counter is reset.) If any request fails, the circuit breaker assumes that the fault is still present, so it reverts back to the open state and restarts the timeout timer to give the system a further period of time to recover from the failure.

We will implement the retry pattern in conjunction with the circuit breaker pattern in the listener service. Because we will mostly perform HTTP operations, we will create a separate class and implement these patterns to provide resiliency.

In .NET, there is a very popular library called the Polly framework (see *https://github.com/App-vNext/Polly*) that can be used to implement retry and circuit breaker patterns. First, we will first install the Polly library from *NuGet.org* into our listener service project.

In Listing 6-5, we used the IResilientHttpClient base class and expose methods such as Get, Post, Put, and Delete:

LISTING 6-5 IResilientHttpClient interface

```
namespace EventListener
{
    public interface IResilientHttpClient
    {
        HttpResponseMessage Get(string uri, string authToken = null);
        HttpResponseMessage Post<T>(string uri, T item, string authToken = null);
        HttpResponseMessage Delete(string uri, string authToken = null);
        HttpResponseMessage Put<T>(string uri, T item, string authToken = null);
    }
}
```

We then extended this and implemented it to create the following ResilientHttpsClient.cs class. You can see the full code on the Microsoft Press website for this book, but we want to call out a few particular parts of the code shown in Listing 6-6:

LISTING 6-6 Circuit breaker and Retry patterns

```
//Circuit breaker policy: if 2 exceptions occurred, it will break the circuit (open the
circuit) for 10 seconds.
  _circuitBreakerPolicy = Policy.Handle<AggregateException>(x =>
        {
            var result = x.InnerException is HttpRequestException;
            System.Console.WriteLine("Circuit opened...");
            return result;
        })
        .CircuitBreaker(exceptionsAllowedBeforeBreaking: 2, durationOfBreak: TimeSpan.
         FromSeconds(10));
//Retry policy: It will continuously retry
        _retryPolicy = Policy.Handle<AggregateException>(x =>
        {
            var result = x.InnerException is HttpRequestException;
            return result;
        }).RetryForever(ex => System.Console.WriteLine("Retrying..."));
```

Because each of these methods use both retry and circuit breaker patterns internally, it helps us to achieve resiliency. Listing 6-7 shows the complete code for the ResilientHttpClient class:

LISTING 6-7 Complete ResilientHttpClient Class

```
using Microsoft.AspNetCore.Http;
using Polly;
using System;
using System.Collections.Generic;
using System.Linq;
using System.Net;
using System.Net.Http;
using System.Net.Http.Headers;
using System.Threading.Tasks;

namespace EventListener
{
    public class ResilientHttpClient : IResilientHttpClient
    {
        static Policy _circuitBreakerPolicy;
        static Policy _retryPolicy;
        HttpClient _client;
        IHttpContextAccessor _httpContextAccessor;
        public ResilientHttpClient(IHttpContextAccessor
httpContextAccessor, HttpClient httpClient)
```

```csharp
        {
            _httpContextAccessor = httpContextAccessor;
            _client = httpClient;
            _client.DefaultRequestHeaders.Accept.Clear();
            _client.DefaultRequestHeaders.Accept.Add(new MediaTypeWithQualityHeaderValue
            ("application/json"));
            _circuitBreakerPolicy = Policy.Handle<AggregateException>(x =>
            {
                var result = x.InnerException is HttpRequestException;
                System.Console.WriteLine("Circuit opened...");
                return result;
            })
            .CircuitBreaker(exceptionsAllowedBeforeBreaking: 2, durationOfBreak: TimeSpan.
             FromSeconds(10));
            _retryPolicy = Policy.Handle<AggregateException>(x =>
            {
                var result = x.InnerException is HttpRequestException;
                return result;
            }).RetryForever(ex => System.Console.WriteLine("Retrying..."));
        }
        private void SetAuthHeader(HttpRequestMessage requestMessage)
        {
            var authorizationHeader = _httpContextAccessor.HttpContext.Request.
            Headers["Authorization"];
            if (!string.IsNullOrEmpty(authorizationHeader))
            {
                requestMessage.Headers.Add("Authorization", new List<string>()
                { authorizationHeader });
            }
        }
        public HttpResponseMessage Get(string uri, string authToken = null)
        {
            return ExecuteWithRetryandCircuitBreaker(uri, authToken, () =>
            {
                var requestMessage = new HttpRequestMessage(HttpMethod.Get, uri);
                if (authToken != null)
                {
                    requestMessage.Headers.Authorization = new AuthenticationHeaderValue
                    ("bearer", authToken);
                }
                var response = _client.SendAsync(requestMessage).Result;
                if (response.StatusCode == HttpStatusCode.InternalServerError)
                {
                    throw new HttpRequestException();
                }
                return response;
            });
```

```csharp
        }
        public HttpResponseMessage Post<T>(string uri, T item, string authToken = null)
        {
            return ExecuteWithRetryandCircuitBreaker(uri, authToken, () =>
            {
                var requestMessage = new HttpRequestMessage(HttpMethod.Post, uri);
                requestMessage.Content = item as HttpContent;
                var response = _client.SendAsync(requestMessage).Result;
                if (response.StatusCode == HttpStatusCode.InternalServerError)
                {
                    throw new HttpRequestException();
                }
                return response;
            });
        }
        public HttpResponseMessage Put<T>(string uri, T item, string authToken = null)
        {
            return ExecuteWithRetryandCircuitBreaker(uri, authToken, () =>
            {
                var requestMessage = new HttpRequestMessage(HttpMethod.Put, uri);
                requestMessage.Content = item as HttpContent;
                var response = _client.SendAsync(requestMessage).Result;
                if (response.StatusCode == HttpStatusCode.InternalServerError)
                {
                    throw new HttpRequestException();
                    Logger.Instance.LogMessage("Error occured " + response.StatusCode);
                }
                return response;
            });
        }
        public HttpResponseMessage Delete(string uri, string authToken = null)
        {
            return ExecuteWithRetryandCircuitBreaker(uri, authToken, () =>
            {
                var requestMessage = new HttpRequestMessage(HttpMethod.Delete, uri);
                SetAuthHeader(requestMessage);
                if (authToken != null)
                {
                    requestMessage.Headers.Authorization = new AuthenticationHeaderValue
                    ("bearer", authToken);
                }
                var response = _client.SendAsync(requestMessage).Result;
                if (response.StatusCode == HttpStatusCode.InternalServerError)
                {
                    throw new HttpRequestException();
                }
                return response;
```

```
        });
    }
    public HttpResponseMessage ExecuteWithRetryandCircuitBreaker(string uri,
    string authToken, Func<HttpResponseMessage> func)
    {
        var res = _retryPolicy.Wrap(_circuitBreakerPolicy).Execute(() => func());
        return res;
    }
  }
}
```

Now that the listener service is complete from the code standpoint, let's make sure that we have it running on a WebJob.

> **NOTE** WebJobs
>
> A WebJob is a feature of Azure App Service that enables you to run a program or script in the same context as a web app. We will be hosting this WebJob inside the Online Auction Application Azure App Service space. For more information on WebJobs, see *https://docs.microsoft.com/en-us/azure/app-service/webjobs-create.*

We packaged the WebJob as a .zip file, uploaded it, and then published it using the following command:

```
dotnet publish -c Release
```

Then you will publish the binaries from the .zip file in the **WebJob** blade in the Azure portal.

Building the Kafka HTTP client

We've now set up most of our prerequisites. Because we have a resilient HTTP Client, we will now create a new KafkaHttpClient class that exposes a method to call the Auction API. Listing 6-8 shows the relevant code snippet for the KafkaHttpClient:

LISTING 6-8 KafkaHttpClient Class

```
public class KafkaHttpClient
  {
      IResilientHttpClient _client;
      IConfigurationRoot _configuration;
      public KafkaHttpClient(IResilientHttpClient client, IConfigurationRoot
      configuration)
      {
          _client = client;
          _configuration = configuration;
      }
      public void UpdateAuctionForBid(string jsonObject)
      {
```

```
        Logger.Instance.LogMessage($"Inside UpdateAuctionForBid where jsonObject
        is {jsonObject}");
        var content = new StringContent(jsonObject, Encoding.UTF8, "application/json");
        HttpResponseMessage response = _client.Put<StringContent>(_configuration
        ["AuctionServiceURL"], content);
        response.EnsureSuccessStatusCode();
        Logger.Instance.LogMessage($"Inside UpdateAuctionForBid, updated table");
    }
}
```

As you can see, we are leveraging our resilient client and then creating the UpdateAuction-
ForBid method, which is the primary function that will be called within the Kafka Consumer for
performing the end-to-end action of updating the auction service with a new Bid. When referring to the KafkaConsumer class, we placed these bold lines into the code (see Listing 6-9):

LISTING 6-9 Completing the KafkaConsumer Class

```
try
                    {
                        var cr = consumer.Consume(cts.Token);
                        Logger.Instance.LogMessage($"Got one message value is
                        {cr.Value}");
                        dynamic data = JObject.Parse(cr.Value);
                        string bidId = data.bidid;
                        string customerId = data.userId;
                        decimal bidAmount = Convert.ToDecimal(data.bidAmount);
                        int auctionId = Convert.ToInt32(data.auctionId);
                        Logger.Instance.LogMessage($"Calling Update Auction
                        for Bid for {cr.Value}");
                        _client.UpdateAuctionForBid(cr.Value);
                        Logger.Instance.LogMessage($"Consumed message '{cr.Value}'
                        at: '{cr.TopicPartitionOffset}'.");
                    }
catch (ConsumeException e)
                    {
                        Logger.Instance.LogMessage($"Error occured: {e.Message}
                        Stack Trace: {e.StackTrace}");
                    }
```

As mentioned before, we are in the while loop that checks to see when Kafka messages are
coming into our topic. These lines above are added to our try-catch construct that will now
leverage the UpdateAuctionForBid method and subsequently complete the architectural flow
to the auction service.

Summary

In this chapter, we covered:

- Various types of communication practices and methods:
  - The differences between synchronous and asynchronous communications
  - The differences between request/response and pub/sub methodologies
  - Best practices for communication in a microservices architecture
- Pub/sub communication technologies, such as Kafka and RabbitMQ, as well as integrations with some Azure technologies, such as Azure Event Hubs
- How to set up Kafka in the OAS as a reference for setting up your own communication patterns in your microservices architecture:
  - Creating a KafkaProducer and KakfaConsumer
  - Building a hosted listener service and a resilient HTTP client
  - Understanding resiliency and related patterns and how to implement them

In the next chapter, we will cover key security principles and use Azure AD B2C for user authentication and authorization scenarios. We will show how you can set up a new Azure AD B2C tenant, how to create user flows, customizations you can make, and integrations with existing applications.

Chapter 7

Security in microservices

In this chapter, you will :

- Cover key identity and security features related to microservices and application architecture

- Learn about Azure AD B2C and how it can be used for authentication and authorization scenarios

- Walk through the OAS security features and learn about B2C tenant creation, creating user flows, customization, and integration

Security is a topic that is at the forefront of every business today. This is especially the case for the enterprise technology groups within those companies and industries because a lapse in security can lead to reputational damage, fiscal damage, and much more. In today's environment, many businesses rely on their customers to trust them with sensitive data, information, and more. Making sure that there is a security strategy in place for addressing security at each layer is important, especially when we are thinking about app and workload security, as many times these functions are the drivers of the business itself.

In this chapter, we will begin by going over the basics of security in the paradigm of microservices architectures. Then we will go into detail regarding Azure B2C, which is one of the major tools we have used in the case study we have built in this book, and then we walk through steps to set up the OAS.

An overview of security and architectures

Security is a component found in many layers within a company's complete IT structure. Entire books have been written about the intricacies of the OSI (Open Systems Interconnection) model. That being said, this chapter will focus on security that is specifically related to application architectures and microservices to address security considerations that you might see when building these yourself or in other architectural paradigms. It is vital that you understand both the theory and the experiential way of addressing the security of your applications. Our demonstrations with the OAS will arm you with the necessary understanding.

IaaS and PaaS architecture security

An important component of the overall security of the microservices architecture is understanding the individual component security. As we mentioned earlier, we leveraged both IaaS
and PaaS to build out the OAS. Understanding the inherent security of each will help us maintain the holistic security.

IaaS security

When we discuss IaaS, virtual machines are the first services that come to mind. Although we
actually use an app service within the OAS, it is reasonable to say that a collection of VMs could
be the hosts of running multiple microservices within an example environment. That being
said, VMs—especially from the standpoint of IaaS—have many security capabilities, including
default and pre-defined network security; access control through the use of identity; constant
updates and patches from the provider itself; resiliency through the use of availability sets and
zones; and even encryption of related VM resources, such as disks.

Keeping your architectural components up to date is important, as is network security,
because you want to be careful about the VMs' access abilities, as well as for the overall posture for your IaaS components. There are many tools that can help assist with this, but Azure
Security Center is, first and foremost, a tool to help improve the security posture of your on-
premises or cloud environment. Thus, there are many advantages to using it, including access
to a recommendations engine that provides guidance for further bolstering your network
security by performing actions that have quantitative scores associated with them. This gami-
fies security by introducing a Secure Score, which attempts to rate your environment's security
and recommend actions to boost that score. For example, the just-in-time (JIT) VM access
enablement can boost your Secure Score. Also, ensuring that DDoS (distributed denial-of-
service) protection is available can prevent potential breaches.

Infrastructure-as-Code security principles

Although Infrastructure-as-Code (IaC) doesn't immediately elicit thoughts of security, it is important to point out how we can replicate our application architecture in case we accidentally delete infrastructure resources or even if we have to spin up a new instance of our application in another environment. Regardless, it is useful to have a procedure for repeatedly deploying IaaS and even PaaS resources for a given architecture.

ARM (Azure Resource Manager) templates are the tool of choice when deploying Azure resources such as container registries, app service plans, application insights, databases, and much more. You can define these within an ARM template and run them to automatically create the needed Azure resources in your tenant.

Using ARM templates that can deploy repeatable resources for a logical unit is an IaaS best practice. To help picture this, let's draw a parallel to our OAS. Within our application, we have deployed a database for each service. Following this same pattern, ARM templates can be set with variables so that at runtime, you can define the actual database server you will need to be deployed for each resource group that can stand in as the logical unit for each service. We can continue to build on that by instantiating resource SKUs by simply clarifying those parameters and then running the templates.

Thus, we can set up all the microservices back-ends in a very structured way. If you need to build the application with new services that follow this patterning, you can automate this process by making small changes. You can go even a step further by combining app insights for particular service plans, having agents installed to address disk encryption, collecting log information, and more.

Defining these add-ons—whether part of the Azure portal or within templates—is a primary point for the security of your architecture because when it comes to defining monitoring agents or inserting disk encryption, this all contributes to the safety of your IaaS components.

This automation lifecycle will be further discussed in Chapter 9, "Build and deploy microservices."

PaaS security

Primarily, we have used and benefited from many PaaS services within our applications. PaaS acts as a cloud service provider (CSP) platform, which includes many security benefits that are already tied in with services that include container registries, app service plans, Kubernetes, databases, and more. At the same time, PaaS doesn't cover all security concerns, and securing parts of the stack is your responsibility. See Figure 7-1.

Figure 7-1 demonstrates the ownership breakdown between a CSP and the customer. SaaS covers all services; PaaS covers all services except hosted apps; IaaS covers storage, network security and datacenter, operations, and service.

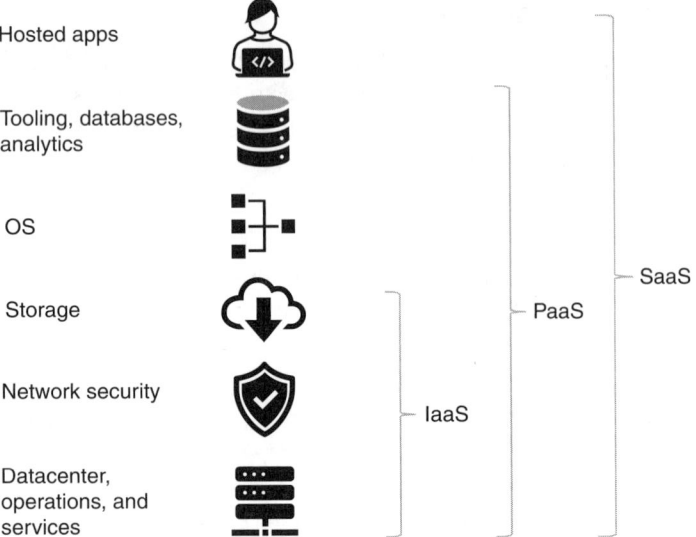

Hosted apps			
Tooling, databases, analytics			SaaS
OS		PaaS	
Storage			
Network security	IaaS		
Datacenter, operations, and services			

FIGURE 7-1 SaaS, PaaS, and IaaS technologies

In our examples of building out application architectures, you will be responsible for iden-tity infrastructure, application-based security (secure-by-code principles), and networking. Again, there are best practices to help minimize the risk of taking on these tasks yourself.

For example, using API Gateways and endpoint security is extremely important and can be provided with PaaS services as well; we will go into further detail on this in Chapter 8. Of course, there are more PaaS security details than those listed in this chapter, but we wanted to highlight some of the most common concerns that you will need to address in your environment.

Azure Storage accounts and key management

Storage accounts can store data of various types and sizes. You can find more detail about Azure Storage Accounts in the Microsoft documentation, but in this chapter, we will focus on the security aspects. Overall, you need to be structured in the separation of resources, and you should have specific storage accounts for VMs, logging and diagnostic info (especially for key management), and even notes for your deployments with IaC.

When you are sharing the information included within the storage account and dealing with related access control, key management comes into play in conjunction with some of the native encryption capabilities of the service. Azure Key Vault (a PaaS service) can maintain keys and secrets for the functionality of particular components or even applications. This provides even more security in conjunction with restricted access to those keys, and it provides strong governance and compliance principles with key rollovers and disk encryption with keys. Over-all, these two platforms have rich feature sets that will serve you well within a microservices architecture because they are well secured.

Databases

Databases will be core to many microservices architectures and thus, their security is highly important. For PaaS databases like Azure SQL Database, security features are already in place, including a built-in firewall with network rules, authentication (AuthN) schemes, granular data entry encryption, and integration with other Azure PaaS services and security products that bolster its security posture. This was a defining reason for us to demonstrate how one of our services leverages the Azure SQL Database and is a great way to create secure database instances.

App services

App services are important PaaS services because they give you the ability to create web/mobile applications for multiple on-premises or cloud devices. Azure's version of this PaaS service is known as Azure App Service which we built out within the OAS. In terms of security best practices, we have used top-notch identity frameworks such as OAuth 2.0 and modern authentication, integrating it with role-based access control for enforcing the principle of least privilege, and we even have some network security baked in with VPN integrations that restrict specific incoming IP Addresses. We will dive deeper into more of these topics in the next section.

> **NOTE** Details on container registries and AKS are covered in more detail in Chapter 9, "Build and deploy microservices."

Zero-trust architecture

Recently, Microsoft has been evangelizing a concept called the zero-trust architecture. The zero-trust architecture is an important cybersecurity framework that builds on many of the principles and guidelines we follow today with NIST, ISO, and so on. The main takeaway is to not trust unmanaged devices, users, applications, software, and so on. Regardless of the medium, you should always verify!

> **NEED MORE REVIEW?** Zero-trust architecture
>
> You can learn more about this concept at *https://www.microsoft.com/en-us/itshowcase/ implementing-a-zero-trust-security-model-at-microsoft*.

For this framework to have any chance of working, authentication and authorization must be key. Although these are technical identity terms, in their simplest sense, authentication means a check is made to see whether you are who you say you are, and authorization checks to see whether you have access to said medium. These are the main tenets of this framework, a trend we are increasingly seeing being adopted by enterprises around the globe. In fact, a commonly used phrase in the industry now is: "Identity is the new perimeter," in which identity is a primary frontline for our security frameworks. We are able to build out many components of this framework with Azure AD, which we will show later in this chapter.

Aside from identity, another tenet of the zero-trust model is to ensure the security and health of your devices. This primarily applies to instances in your environment where you could have unmanaged devices. Having Mobile Device Management and Mobile Application Management (MDM and MAM) tools such as Microsoft Intune will help address this gap, contributing to the zero-trust and identity story. Let's dive into these topics.

Identity is the new perimeter

The "identity is the new perimeter" concept isn't entirely new or unheard of, but its implementation is important when establishing a strong first line of security.

When thinking about the word "perimeter," we have to harken back to its original context as a physical network perimeter in much older, traditional enterprise environments. In the past, having this physical network perimeter was the norm. Resources were locked down and kept in a single area where nothing can go in and out of this impermeable structure. However, times change, and today, you don't even have to be at your workplace to get work done thanks to VPN connections, bring your own device (BYOD), and the like that allow many to work from home. Of course, this means the previously isolated perimeter is now porous and malleable, and we've had to start innovating in the identity space to account for these situations. The workforce can now access the tools and resources that they need from anywhere and any device/platform.

As we move toward these more secure models of building applications, IT teams must make use of many identity management and access-control technologies such as Azure Active Directory to maintain control over corporate environments.

Azure Active Directory

A great best practice within any application architecture is to use strong, modern authentication and authorization. Though there are many tools available, including Ping, Okta, ADFS, and so on, we used Azure AD B2C for the OAS. No matter the technology you use, you must design your authentication and authorization flows in relation to the diversity of users and how they interact with your application.

Azure AD allows us to use federated identities and can work in conjunction with on-premises Active Directory or other directory stores to extend synced identities to the cloud. This is used in conjunction with libraries such as ADAL and MSAL that provide best practices in the form of code to build your authentication and authorization flows, generating a more secure user experience. In addition, technologies such as multifactor authentication and modern authentication protocols bolster security further and create a strong first line of defense.

Also, Azure AD allows you further coordinate with partners and consumers who leverage user flows for specific use cases and scenarios that still leverage the overall strength of Azure AD.

For example, let's say that we have an enterprise application that was built on a microservices architecture for Contoso. Let's also posit that this is an internal user application. If we have a requirement where internal users need to be logged in seamlessly, benefits like SSO would make a world of difference to their access experiences. Thus, there is a clear benefit of registering the application on Azure AD to use modern authentication (which you should because it's a best practice) as well as configurations to set up SSO and more.

If we had a scenario in which we had external partners accessing this application for some hypothetical use case, then we could use Azure AD for their needs as well. Some options include adding partners as guests into your Azure AD tenant, setting up federation between their directory and Contoso's, and developing other advanced flows to provide a more seamless access experience. As for access control, roles and permissions can be set to ensure access to particular parts of the application for maintenance. This separation of duties and privileges is important and will be discussed next.

Role-based access control (RBAC) and the principle of least privilege

Access Control is a very important concept in relation to identity. Making sure certain individuals have access to the resources at certain levels is important to ensuring the application stays secure, especially when dealing with PaaS services in your architectures. But the question remains, how do we determine who should actually get access?

This gives way to the principle of least-privilege concept , which dictates that user access is specifically outlined for each user to have just enough to perform his or her job functions. This prevents people from having unneeded access, which deters insider risk scenarios, lateral threat scenarios, and more.

Having this principle is great, but how do we enforce it? Azure gives us role-based access control (RBAC), which allows us to provide the appropriate access needed for basic functions to users, groups, and applications. You can use built-in or custom Azure roles that you can create with JSON with particular permissions and rights, with the ultimate intent of assigning objects to those roles that define their privileges. Of course, when a user needs occasional access to higher privileged abilities, technologies such as privileged identity management grant that access within certain limitations.

> **NEED MORE REVIEW?** Azure RBAC
>
> For more information about RBAC, see *https://docs.microsoft.com/en-us/azure/role-based-access-control/overview.*

Managed identities

If you are considering building out applications to be used on the enterprise level, the concept of managed identities will come into play. This streamlines authentication by separating credentials from code. When used in conjunction with key management services such as Key Vault, we have a more secure way to access resources, such as virtual machines, databases, storage accounts, and more, ultimately avoiding antiquated and difficult-to-manage service principals.

Multifactor authentication

One of the most common secure practices that we see for access control within enterprise applications is the use of two-factor authentication or the more advanced version, multifactor authentication (MFA). This is a process in which we use several indicators to verify whether "you are who you say you are." These factors include a phone call to verify that you are the requestor; an authentication code through an app or a PIN being sent to a secured account that is only accessible by you. MFA is becoming much more common as a means of ensuring that privileged accounts remain secure. Although MFA is useful, other facets should also be in place for bolstering this security. Technologies such as conditional access consider factors such as location, device health, user risk, app risk, and more at the time of authentication. This creates an authentication flow that is secured from the standpoint of "what you know, what you have, and what you are," and it considers external factors as a determination to see whether a user should be granted access.

Authentication and authorization flows

Before getting into the details of particular authentication flows, let's discuss some baseline knowledge on some of the modern identity protocols. One such example is the OAuth 2.0 authorization framework. According to the specifications written by the Internet Engineering Task Force (IETF) this allows third-party apps to get limited access to HTTP services in one of two ways: Either the access is secured on behalf of the resource owner by moderating approval between the HTTP service and the resource owner, or access is granted when a third-party app obtains access on its own.

> **NEED MORE REVIEW?** More on OAuth 2.0 Authorization Framework
>
> **For more information about the OAuth 2.0 Authorization Framework, see**
> *https://tools.ietf.org/html/rfc6749.*

What does this enforce? With this protocol, you can now trust applications to store your credentials securely and not misuse them. How is this constructed? When thinking of an authorization flow such as OAuth 2.0, you want to make sure that you have both a target resource to access within an application and an Authorization (AuthZ) server in place. The former will be registered to the latter. This could be done with tools such as Azure Active Directory or other tools.

OpenID Connect, also known as OIDC, is a protocol that builds on OAuth 2.0 by gathering information to verify the user, which mostly mitigates the fact that OAuth solely uses the AuthZ protocol. This means that if the flow to acquire an access token is taken away, what we have instead is a strong AuthN flow. This is done using a JWT token, which is an ID token to contain a set of claims to verify a user. Ultimately, our goal is to ensure that these modern protocols are used with the right flows in order to ensure the best security possible, but that raises the question of which flows are currently being used.

First off, why do different flows even exist? With any of these modern protocols, the token sits at the center of its functionality, thus attempts will inevitably be made to steal, hijack, and use them for replay attacks. Even misconfigurations can result in breaking down your own access control.

What kinds of factors do these flows have to provide the optionality needed to have a secure application? Options dependent on client type, token type, and even user accessibility are important distinguishing factors for these flows. Thus, it is important to know these flows from the standpoint of developing your application as your requirements will dictate what protocol you will use, what flow you will use, how users will interact with the application, and in what medium users will interact with it. Having at least some information regarding this optionality will help you develop a more secure microservices architecture.

Authorization code grant flow

One of the several types of flows is the authorization code grant flow. This actually uses a confidential client where client secrets and a direct route to the authentication server exists for greater protection. Now, there are two types of Auth code grant flow, and the second uses a public client. This is what several refer to as Auth Code with PKCE (Proof Key for Code Exchange) flow as the client secret is removed from any potential interceptions and instead uses a hash code. A frequent use case of this is with mobile applications due to the numerous applications that can exist on a singular client, providing better mobile app security. You can see how this flow looks below in Figure 7-2.

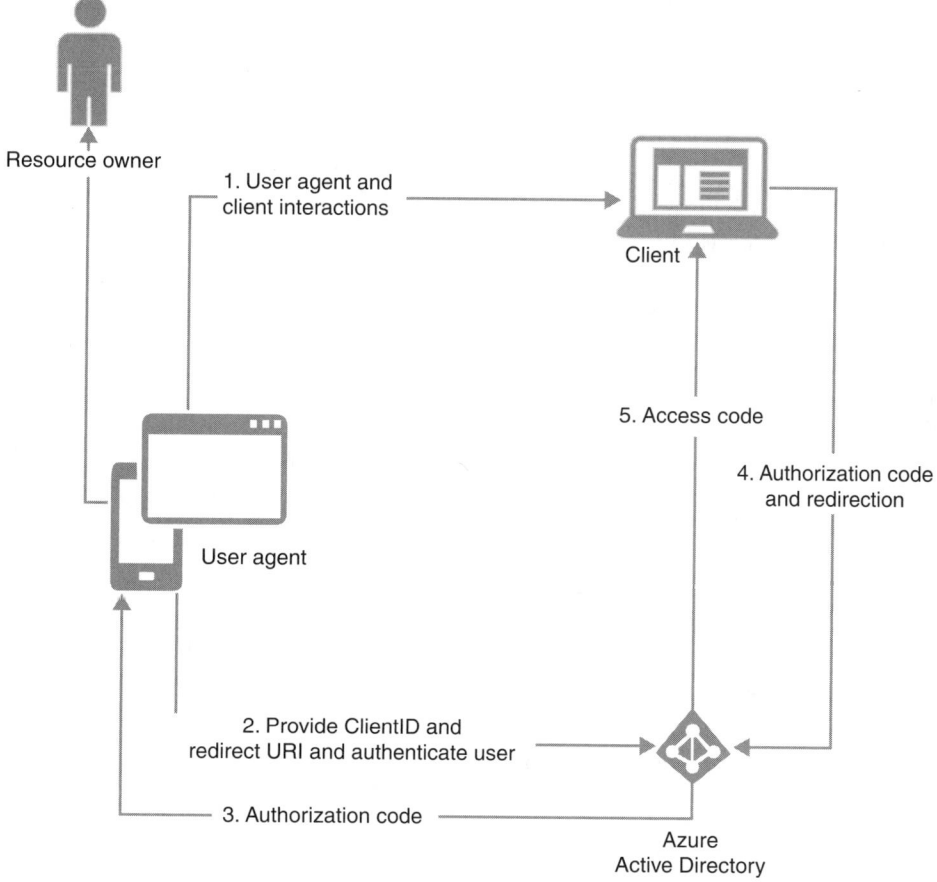

FIGURE 7-2 Authorization code grant flow

In Figure 7-2, we start with the User Agent, which interacts with the client and then provides the ClientID and redirect URI to the authorization server, along with authenticating the user. This results in gaining the authorization code, which is passed to the user agent, which then interacts with the client where the authorization code and redirection is provided to the authorization server. We now have the access code to operate the application.

NOTE Authorization Code Grant

Figure 7-2 is based on the Authorization Code Grant flow diagram from the IETF at *https://tools.ietf.org/html/rfc6749#section-4.1.*

Implicit grant flow

Implicit grant flow has particular use cases oriented toward browser-based clients, so it is best integrated with single-page applications (SPAs). The peculiarity with this particular flow is the maintenance of the state. As such, this flow offers the ability to redirect fragments that help maintain the state throughout the entire lifecycle of the flow. In fact, this is what we leveraged for the OAS. We will provide details about how to configure this flow later in this chapter. See Figure 7-3.

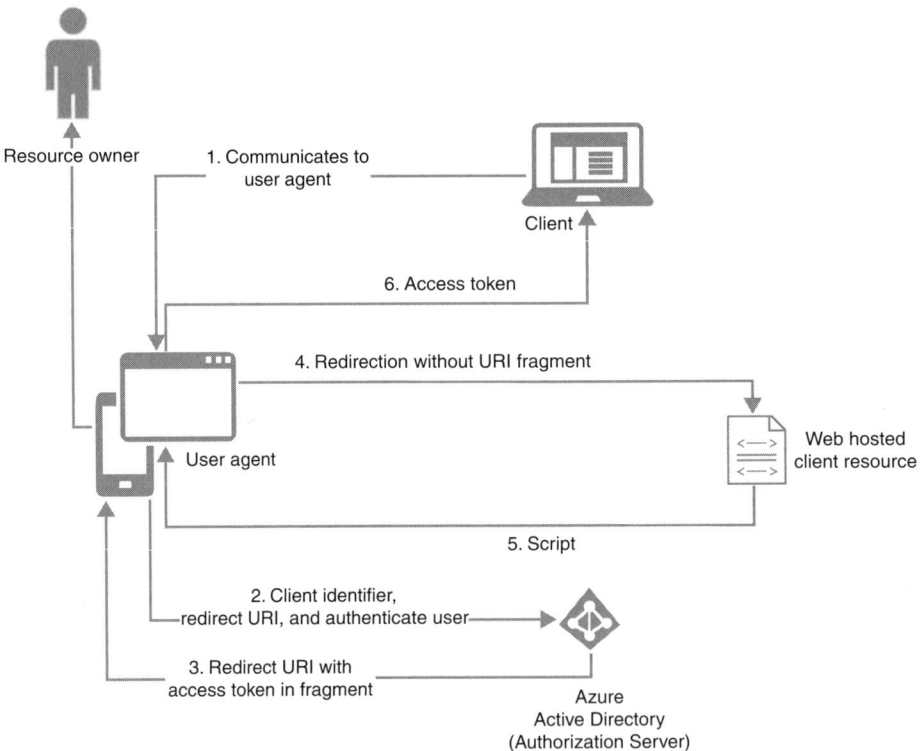

FIGURE 7-3 Implicit grant flow with sequential steps for the flow

Figure 7-3 demonstrates the implicit grant flow with sequential steps. We start with the client communicating to the user agent. The client identifier then redirects the URI and authenticates the user to the authorization server. The redirect URL and access token in the fragment goes to the user agent where it uses the redirection without the URI fragment to the web-hosted client resource. The web-hosted client resource then provides the script to the user agent and then uses the access token to access the client.

> **NOTE** Implicit grant flow
>
> Figure 7-3 is based on the Implicit grant flow diagram from the IETF at
> *https://tools.ietf.org/html/rfc6749#section-4.2.*

OIDC hybrid flow

The OIDC hybrid flow is an amalgam of both the authorization code grant flow and the implicit grant flow. The benefit is that the identity token is part of the OIDC and can verify the user while simultaneously acquiring an authorization code, as well as being part of a single flow. This leverages confidential clients and enables a more efficient modern flow that is becoming the standard for many modern applications.

Let's compare and contrast these three different flows. First, these flows can have access tokens returned at different endpoints. The authorization code grant flow uses two endpoints: the authorize endpoint and the token endpoint. The authorize endpoint provides you with the code, and the token endpoint provides you with the access token in exchange for the code. With the implicit grant flow, a single request is made to the authorize endpoint to directly obtain the access token. The hybrid flow is similar to the authorization code grant flow because both the code and the access token are obtained from the aforementioned endpoints. However, with the implicit flow, the access token is obtained in a single request. We can actually use refresh tokens with the authorization code grant flow and the hybrid flow, but we cannot use this with the implicit grant flow.

One point to acknowledge here is that we have different response types for various flows. As you know, with the authorization code grant flow, you are simply provided with the "code" that refers to the AuthZ code, which is a part of that flow. For the implicit flow, there is the id_token that stands in for the identity token obtained, and there is the fragment, which is indicated with the id_token token response. Hybrid is a combination of the aforementioned response types including the AuthZ code, Identity token, and even the fragment from the implicit grant flow. Overall, one of the most frequent response types we see as a result is code id_token.

Client credentials grant flow

The client credentials grant flow is used only with confidential clients and is a non-human touch process where the client initiates the authentication process with the authorization server and in return, receives the access token. This is a form of direct authorization that is very specific to strong-credentialed users and devices. You can see how this flow looks in Figure 7-4.

Client

2. Access token

1. Authentication

Azure
Active Directory

FIGURE 7-4 Client credentials grant flow

The flow starts with the client authenticating to Azure Active Directory, obtaining the access token, and then passing it back to the client.

> **NOTE** Client credentials grant flow Figure 7-4 is based on the Client Credentials Grant flow diagram from the IETF at *https://tools.ietf.org/html/rfc6749#section-4.4.*

Device code flow

Device code flow is a manner in which to authenticate and receive authorization for devices or apps that do not have the ability to show a browser, such as Internet-based devices like Internet of Things (IoT) or smart devices. As is the case, we instead have a secondary device that does have a browser as part of the overall flow. The client would first request the token from the authorization server, and in return, the client would receive a short code. This code is again used in conjunction with authentication to the authorization server to receive the access token for the client to use and access the service. You can see how this flow appears in Figure 7-5.

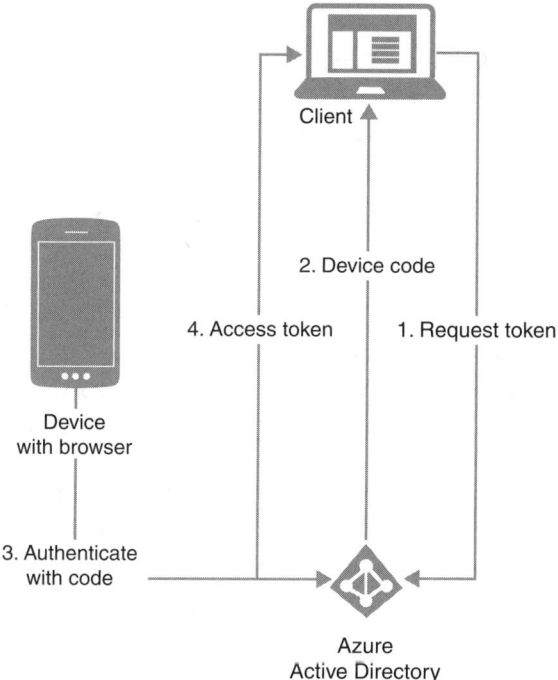

FIGURE 7-5 Device code flow

First, the client requests a token from the authorization server and then receives the device code to the client. With the code, a user can utilize a device with a browser and input the code to authenticate and then receive the access token to access the client.

Azure Active Directory B2C

Now that we have gotten a good baseline on what security looks like in enterprise architectures, let's extend it further by building it out experientially with some background on the primary tool used in the OAS's security service.

We primarily leveraged Azure Active Directory B2C as our primary tool and thus we will use this section to go into some depth about how B2C works, and then we will transition to the implementation of it in the OAS.

The application you build will have some type of target audience. This audience is also what helps determine the functionality of the application and in our context, the kinds of microservices that exist. In fact, the target audience of an application can also impact the types of user flows that exist and need to be configured within the application itself. Let's review some of them in this section.

B2E user flow

B2E can be construed as business-to-enterprise or even business-to-employee. The audience is the overall enterprise environment, meaning the users within your direct organization.

The experience is mostly controlled by the directory's tenant admin (if we were thinking in AAD terms). In the B2E flow, we could have on-premises users who are synced to the cloud or direct cloud users who are provisioned at the direction of the administrator, based on least privilege. This is different than B2C, where you have the ability to directly register yourself for the service.

An example of this flow could be illustrated through HR applications. HR applications have employees as their audience. In cases such as these, the identities of your users are likely to be situated within your environment, whether it be on-cloud or on-premises and most likely in some directory store with associated attributes. To gain access to the corporate application, these identities are likely authenticated against the AuthN/AuthZ server through the flows explained earlier in this chapter. In this case, the server might use Azure AD and thus, will allow your enterprise users to access the application. Azure AD is just one example, of course, but this example helps illustrate a corporate app with corporate users and how an identity provider can help address the B2E flow.

B2B user flow

B2B means business-to-business and centers on an audience that is external to the business/enterprise. Likely, the external audience is comprised of your company's business partners. For example, financial services companies tend to work with other organizations for particular services, such as due diligence, mortgage handling, insurance, and the like. Ultimately, these partners and businesses will need access to some of the resources that your company possesses, so providing these partners and businesses with an optimal experience while being secure is vital. B2B user flows cover these needs while enabling use cases, such as file/external sharing. Azure AD B2B helps with the creation of B2B accounts or even allows partner self-management if the partner utilizes similar technology and has its own respective tenants as well.

The B2B user experience could start with the Azure AD tenant admin enabling Azure AD B2B and inviting guest users. If the guest user is part of an existing federated Azure AD tenant, the user will receive the email notification on the access level they now have within the inviter tenant. If the guest user does not have an associated account, a registration and redemption process will take place so that the user can use his or her local email account to create an MSA (personal Microsoft account) and then become an official B2B user with a certain level of access.

Much of the optionality in this area is beyond the scope of even Azure, and though it is important to be productive and share resources and information with company outsiders, it is important that you maintain the integrity of your overall IT environment.

B2C user flow

B2C is generally referred to as business-to-consumer and focuses on audiences that are external to the enterprise. Here, the target audience could range from niches of typical users to consumer groups. B2C is unique in the sense that it is geared toward an audience of generic consumers. For example, the fictional company Contoso might build a social media application so users can log in to interact with other users who have created an account in the same application. Unlike B2B and B2E, there is less control over who these users are, and maintaining their identities presents a daunting overhead as well. Ultimately, B2C user flows leverage the ability for users to create accounts themselves by inputting info or using other identity providers to set a baseline for providing "who they are." The flows then can use this info in tandem with Contoso authorization servers or with partners to verify users and ultimately provide access to the needed resources.

For the OAS, we have an audience who uses the application to bid and sell products. The users are not constrained to any partnerships and collaborations, nor are they related directly to enterprise. This means B2C is the most sensible of these flows to choose from. Also, because we were building out this application with Azure resources, we decided the native B2C provider would be of assistance, so we used Azure AD B2C.

Azure AD B2C is an identity service that addresses a multitude of B2C use cases. You can use a plethora of different identities such as preexisting social media identities, email, and government IDs to substantiate user actions. Ultimately, these actions will help you use leverage the full richness of the identity, and on the back-end side, you can use other customizations that are geared for better analytics, APIs, and even branding. Ultimately, this also provides a better experience for the user and allows you to leverage SSO and modern identity protocols such as OAuth 2.0 and OIDC. The registration and redemption experiences are more straightforward as well, which provide us with the building blocks we need for our OAS.

The OAS includes three services that can be leveraged by consumers: the auction service, the bid service, and the payment service. To access these services and relevant UIs, we must have a registration service set up using Azure AD B2C and demonstrate how it works end-to-end.

End-to-end OAS security implementation

To understand the end-to-end implementation, Figure 7-6 shows a high-level view.

In Figure 7.6, first, the application is registered in Azure Active Directory B2C, and the ClientID and ClientSecret are noted for later configuration. We used MSAL to configure the authentication. Next, the user can register with a username and password and configure the return token. When going through the actual workflow, the user will be challenged for the credentials and then attempt to consume the APIs by calling the APIM endpoint. You can pass the token retrieved from authentication and use this for validation and then pass on the tokens to the APIs to actually utilize.

End-to-end authentication flow

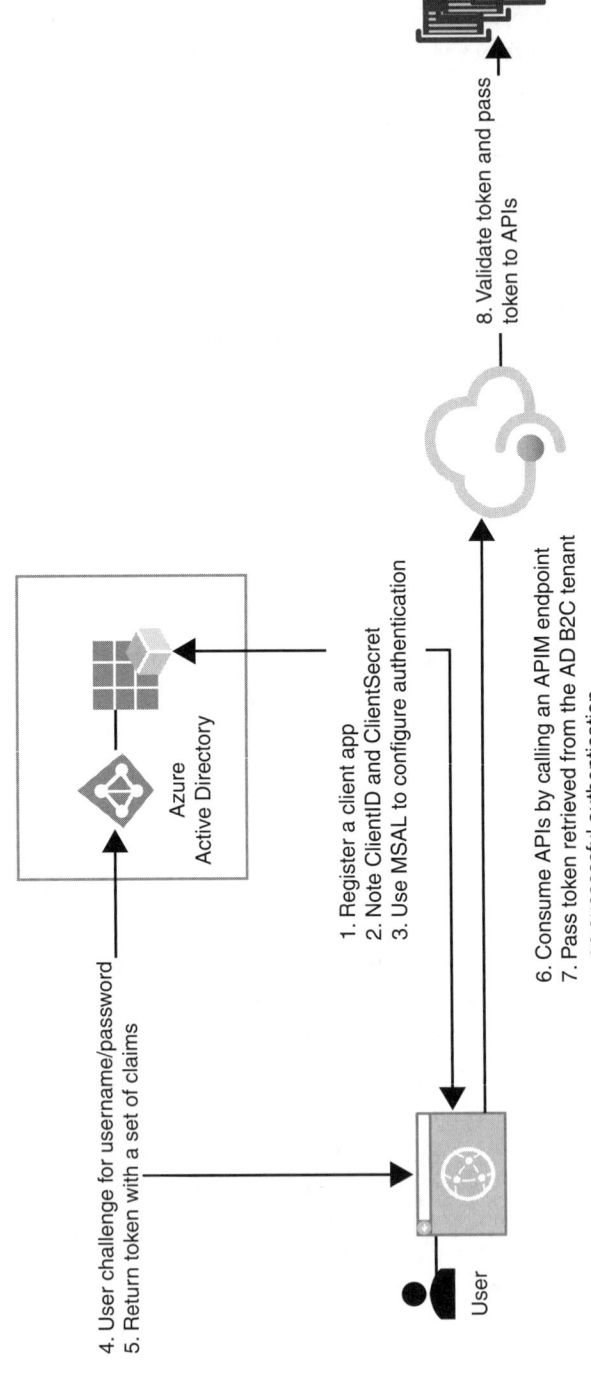

4. User challenge for username/password
5. Return token with a set of claims

Azure
Active Directory

1. Register a client app
2. Note ClientID and ClientSecret
3. Use MSAL to configure authentication

6. Consume APIs by calling an APIM endpoint
7. Pass token retrieved from the AD B2C tenant
   on successful authentication

8. Validate token and pass
   token to APIs

User

FIGURE 7-6 End-to-end workflow for signing in to the Online Auction System

As you can see in Figure 7-6, we attempted to leverage the power of Azure AD B2C by registering a client app and its respective information, such as the ClientID and ClientSecret. Once this information is obtained, we are able to coordinate with the Microsoft Authentication Library (MSAL) and use the ClientID and ClientSecrets to configure our authorization flows. Next, the actual flow begins, where the user signs up for the service with a simple username and password. (In most enterprise examples, you should gather more information for stronger authorization practices, such as automatic enrollment into MFA and so on.) Then the user uses these new credentials to log in to the service and receives a token with a customized set of claims. You can now use the token to consume the back-end APIs by calling the APIM endpoint. All these steps gloss over some complexity, so in the remainder of this chapter, we will go over the instructions to build out a B2C service from end-to-end. We will be doing this via a walkthrough of how we set up the B2C service for the OAS.

User perspective

There is a sign in button at the top right corner of the OAS main page that you are already acquainted with from former chapters. When that is clicked, the user is redirected to the depiction in Figure 7-7. The user can sign in to the portal by entering an **Email Address** and **Password**, and new users can register on the portal by clicking the **Sign Up Now** link.

FIGURE 7-7 User log in page for the OAS

After entering a valid username and password, the user will be authenticated and given access to the services on the OAS main page.

Microsoft Authentication Library (MSAL)

The Microsoft Authentication Library (MSAL) is the modern version of Active Directory Authentication Library (ADAL). Now you can use the Microsoft identity endpoint to get the tokens you need to consume other services such as APIs. In our case, we would specifically use this for coordination with our back-end services after authentication. MSAL provides a very flexible platform to simplify building your own protocol implementation, enable OBO flows, and even coordinate with Azure AD B2C. Some other benefits offered by MSAL include:

- Built-in functionality for OAuth and code libraries
- OBO (on-behalf-of) flows
- Built-in handling for token expiration
- Banners to log in for particular services
- Configuration walkthroughs
- Telemetry and troubleshooting information

Creating a tenant

The first step is to leverage Azure AD B2C by creating a tenant instance. Navigate to the Azure portal and then search for **Azure AD B2C** in the search box at the top. Follow the on-screen steps to create a new tenant, as shown in Figure 7-8.

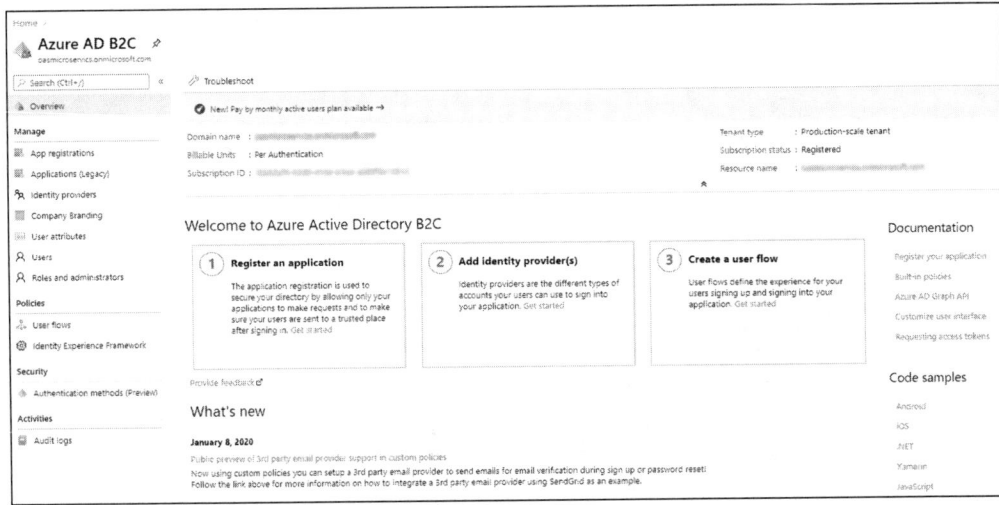

FIGURE 7-8　Main Azure AD B2C tenant page for the OAS

Register your application

First, we need to register our application. In the left-side blade, click **App Registrations**, which will then change the main area of the screen, as shown in Figure 7-9.

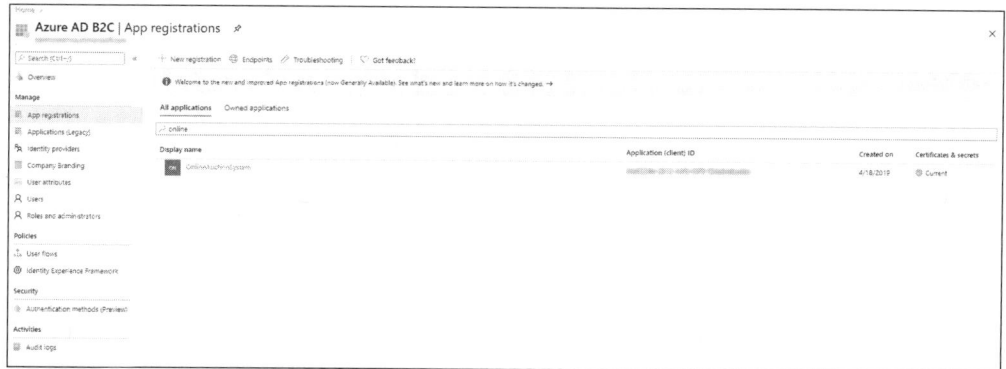

FIGURE 7-9   App Registrations screen for the OAS B2C tenant

Next, let's review the new App Registration just created and take note of the important information for getting certificates and secrets to be current, as well as the overall configuration. Specifically, you can use any text editor to jot down the Application (Client) ID and Directory (Tenant) ID. This, in tandem with the secret, is the basis of your configuration being secure. Also, later configuration steps require you to input this information. See Figure 7-10.

FIGURE 7-10   App Registration

Configuration

Next, click **Authentication** to start setting up the platform configuration, as seen in Figure 7-11.

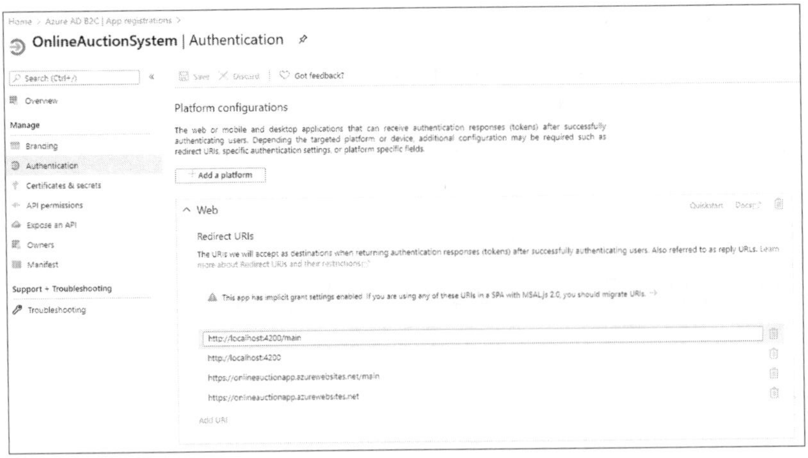

FIGURE 7-11 Authentication pane for the OAS

Specifically, take a look at the **Redirect URIs** section. These are our responses when tokens perform authentication correctly and as you can see, several have already been added. The ones of most consequence are the URIs at these sites:

- *https://onlineauctionapp.azurewebsite.net/main*
- *https://onlineauctionapp.azurewebsite.net*

This is because these are the stand-in responses for our cloud-hosted application and will help us navigate there. You'll notice that there are some localhost options here as well. These were used for testing our local machines before we configured our website to be hosted in the cloud.

Next, see the **Implicit Grant** section, which is where we can provide some configuration details for setting up the implicit grant flow (see Figure 7-12). Because we do invoke web APIs through Angular, we checked both the **Access Tokens** and **ID Tokens** options. Review the instructions for the applications that you are creating and be sure to leverage this tool when possible.

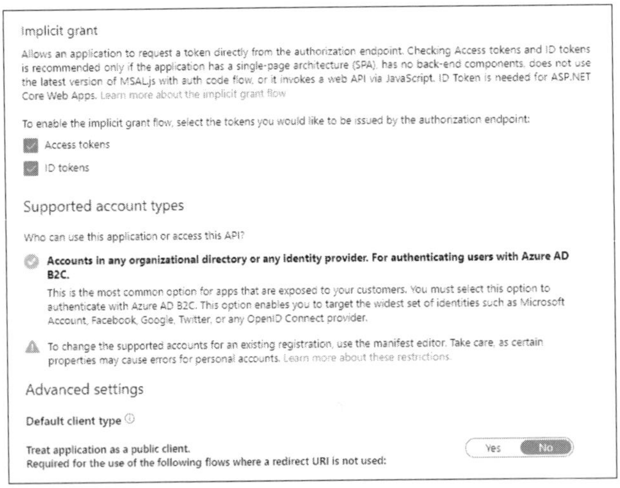

FIGURE 7-12 Implicit Grant and supported account types options

In Figure 7-12, the **Supported account types** option has been left at its default setting—**Accounts in any organizational directory or any identity provider. For authenticating users with Azure AD B2C**—which allows users to leverage Azure AD B2C and several major OIDC providers. This is the only option that can leverage Azure AD B2C, and it provides you the most authentication optionality. Finally, in the Advanced Settings section, the **Treat application as a public client** slider has been set to **No**.

Now that we have reviewed some of the options in the **Authentication** tab, let's move over to the **API permissions** tab, as shown in Figure 7-13.

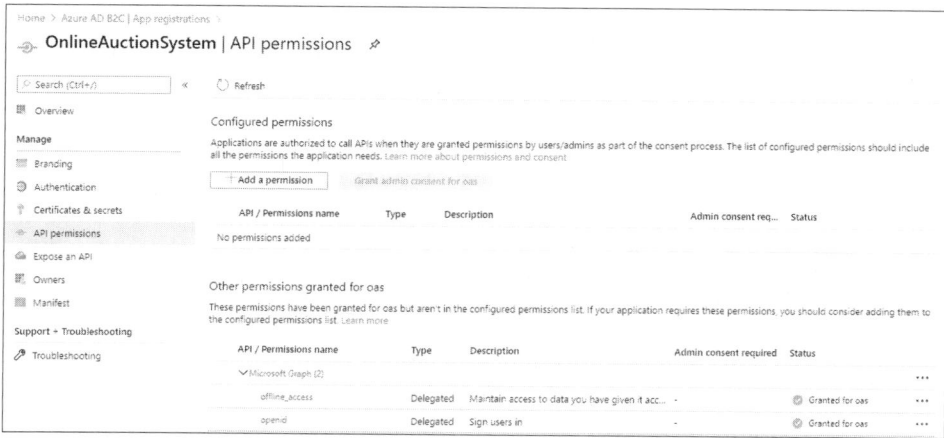

FIGURE 7-13   API Permissions tab for OAS registration

Here, we set some of the permissions to get our OAS to function. Most notably, we used the Microsoft Graph permissions for both offline access as well as for OpenID so that the application allows users to sign in (see Figure 7-14).

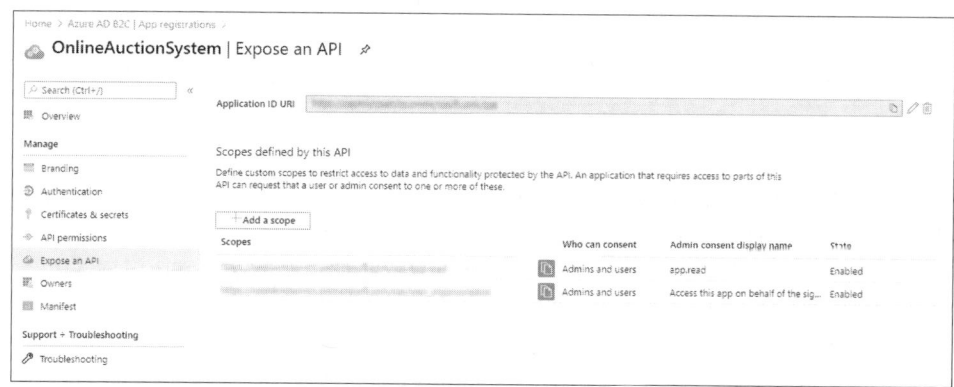

FIGURE 7-14   Expose an API

Next, we have to expose the API to actually coordinate between the OAS back-end services and the B2C authentication service. You want to add scopes in order to provide that ability. We did this by giving read access to the OAS tenant and also an OBO (on-behalf-of) scope to provide the API the necessary delegation and permissions to run noninteractively without a set of user actions.

User Flows

Next, we will set up our actual user flows. On the tenant page that was created for Azure AD B2C, select **User Flows** in the side pane. Here, you will find the screen shown in Figure 7-15, where you have the ability to create user flows. You'll notice that we have already set up three user flows: **Profile Editing**, **Password Reset**, and **Sign Up And Sign In**.

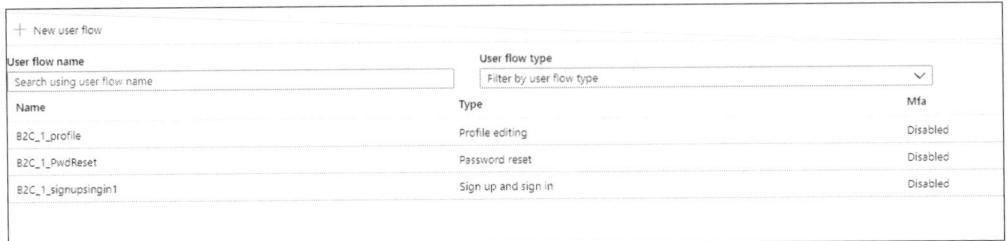

FIGURE 7-15   User flows configured for the OAS

Figure 7-16 takes a closer look at the **Sign Up And Sign In Flow**.

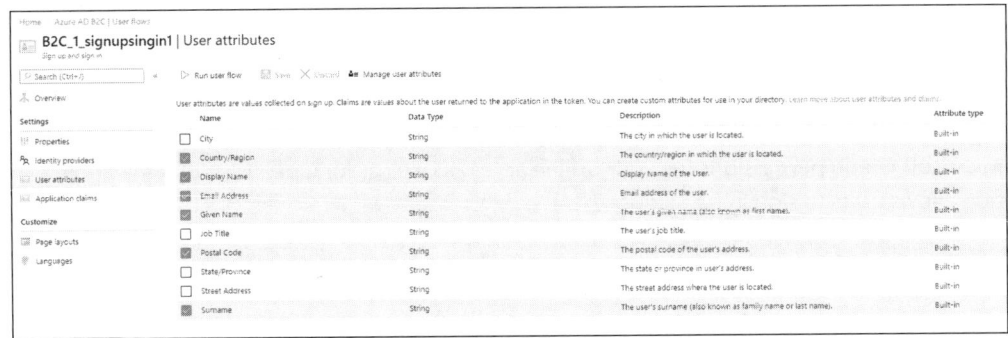

FIGURE 7-16   User attributes for the Sign Up And Sign In flow for the OAS

In the **User attributes** section, you can select the user attributes that you want to collect from those who use this particular service. As you can see in Figure 7-16, several attributes have been selected: **Country/Region**, **Display Name**, **Email Address**, **Given Name**, **Postal Code**, and **Surname**. However, you can customize these as you see fit.

After configuring the options shown in Figure 7-16, you can see the results of your work in the screen shown in Figure 7-17. Here, you see the page for signing up for an OAS account.

FIGURE 7-17 Sign-up and registration page for the Online Auction System

As you can see, the respective fields that were chosen in the **User attributes** selection are shown here on the sign-up and registration page. This is a great way to customize any of your B2C applications with the information that best suits your needs. In fact, this information can even be used within the claims token for the application, as shown in Figure 7-18.

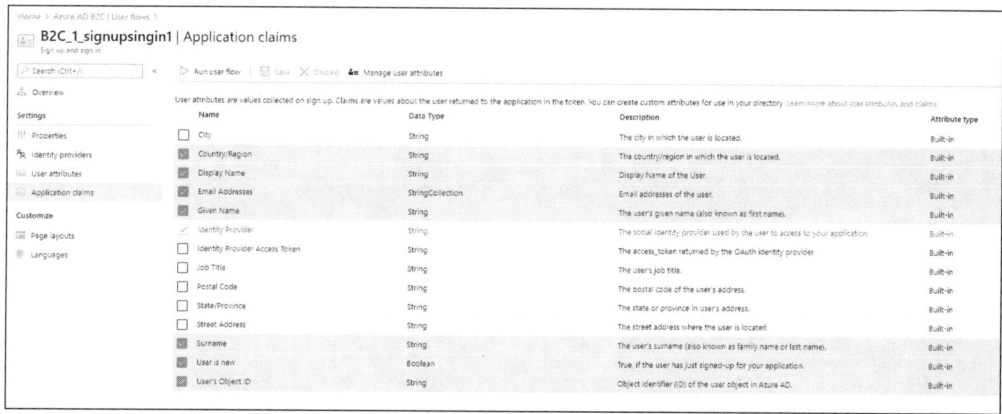

FIGURE 7-18 Application Claims for the Sign Up and Sign In flow for the OAS

Here, you can customize the information you want to include in the application claims, and you can do so by even selecting those that you have configured within the **User attributes** section. We went with some of the built-in options, but again, you can choose these based on your requirements.

In Figure 7-19, the **Identity providers** tab is shown on the left. As explained before, you can select other identity providers to integrate existing user profiles and provide authentication through these resources. For example, we chose to have a local email account be our main identity provider when signing up for an account.

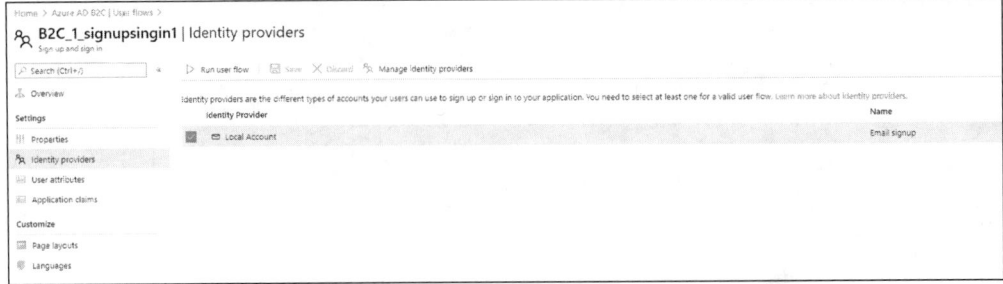

FIGURE 7-19   Identity providers tab for the Sign Up And Sign In flow

After clicking **Manage identity providers** (see Figure 7-20), you will be able to add in some built-in options or create a new OIDC provider. Here, you can see the full range of some built-in identity providers ranging from social media services like Facebook, LinkedIn, and Twitter, all the way to enterprise or consumer accounts such as Microsoft accounts, Amazon, and GitHub. Moreover, you can even set up other options for OpenID Connect providers as well.

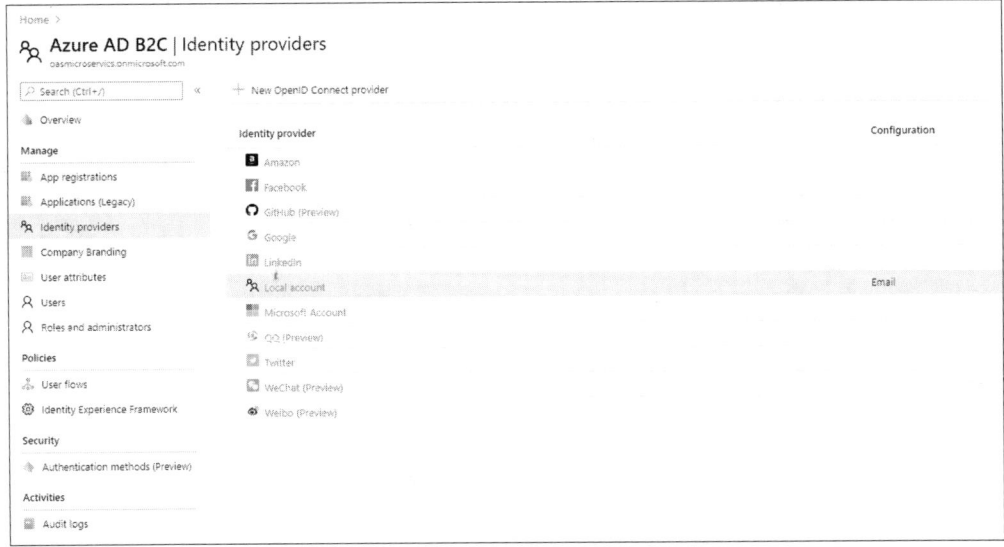

FIGURE 7-20   Identity Providers

Now, in the **User flows** section, you can configure some of the flows that we had set up for the OAS. As you see in Figure 7-21, the **New user flow** tab offers the **Sign up and sign in** flow, which is one of the flows that we set up.

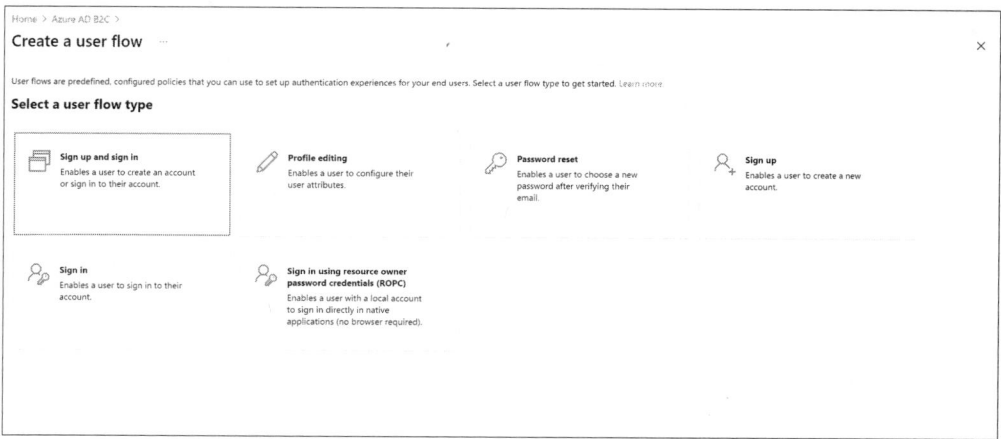

FIGURE 7-21 Create a user flow

Figure 7-22 shows the **Create** screen with the **Sign up and sign in** flow being created.

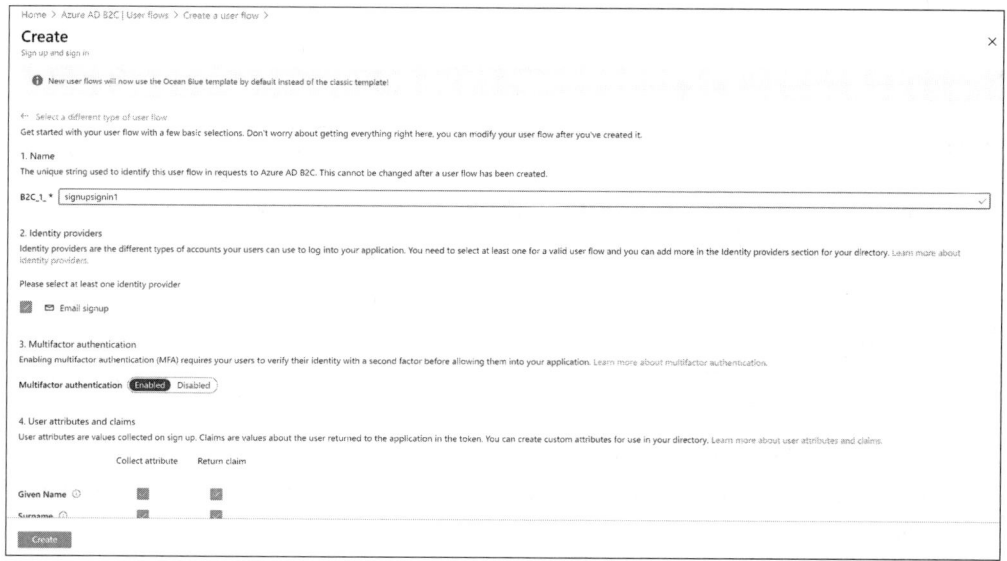

FIGURE 7-22 Creating the Sign Up And Sign In flow

Finally, you can test the user flow to see how it works, as shown in Figure 7-23.

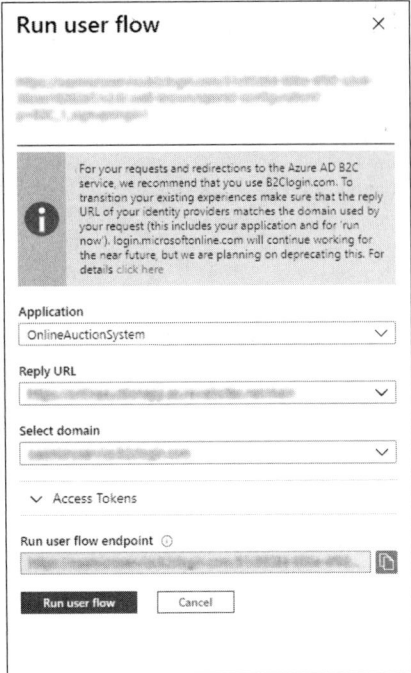

FIGURE 7-23   Run user flow

Ultimately, you can put together all these steps to make cogent flows that bolster your security and can help you secure your applications.

Summary

In this chapter, we covered:

- An overview of security and architecture
- Azure Active Directory B2C and how it can help achieve some of your security goals for your applications
- A complete end-to-end walkthrough on how we implemented security for the OAS through Azure AD B2C

The next chapter will cover the need to establish an API gateway that acts as a single front-door service to our back-end microservices. We will use Azure API Management to enable the API gateway and explore different policies to filter IPs and validate tokens, and we will learn how to enable an authorization server for consumer authentication.

Chapter 8

Set up Azure API Gateway

In this chapter, you will:

- Understand the importance of API gateways in microservices architecture
- Explore Azure API Management and setup and configure API and policies
- Explore Azure API Management for microservices hosted inside Azure Kubernetes Services

Microservices applications are comprised of several services communicating over a lightweight channel. Each service exposes an endpoint that can be consumed by the front-end application. Adding cross-cutting concerns such as security, message transformation, aggregators, IP address restrictions, token validations are some of the key things to consider in securing and exposing microservices. Adding all these features on the service itself can require a monotonous and painstaking effort, and these changes must be applied everywhere when changes are made.

Moreover, adding all the endpoints for each service is not a viable option because any changes to the DNS or IP address require you to update the client application as well. To address these concerns, an API gateway provides a single-layer and front-door proxy to your back-end services and exposes a single DNS that can be consumed by client applications. Moreover, you can harden the security and accommodate all the cross-cutting concerns on the API gateway rather than doing this repetitive work on each microservice.

You can develop a custom gateway or a proxy service to take requests from client applications and then implement cross-cutting features, or you can use any cloud managed API gateway that provides all these features. Some gateway options include Azure API Management, Google Apigee, and IBM API Connect. However, because our solution is based on Azure, we ended up using Azure API Management.

Why do you need an API gateway?

If you don't use an API gateway, your client application needs to communicate directly to the back-end services, which has a few drawbacks:

- The client becomes tightly coupled to back-end microservices.
- The client becomes susceptible to service refactoring/additions.

- The client communication can become chatty, executing network roundtrip calls and incurring latency over mobile networks.

- Microservices are directly exposed to the client, which widens the attack surface.

- The client code becomes complex and cluttered.

- Cross-cutting concerns are duplicated.

The API gateway is one of the most widely used patterns when developing a microservices architecture. It provides a front-end service/endpoint to encapsulate core back-end services. Following are some of the benefits of using an API gateway:

- Exposes a single point of entry.

- Routes requests to back-end services.

- Can circulate a request across multiple back-end services.

- Insulates clients from internal partitioning and refactoring.

- Keeps the gateway close to the back-end services to reduce latency.

- Ensures availability to prevent a single point of failure.

Figure 8-1 depicts three services running inside containers with the web application communicating to the API gateway, which is further routing the traffic to the backend services.

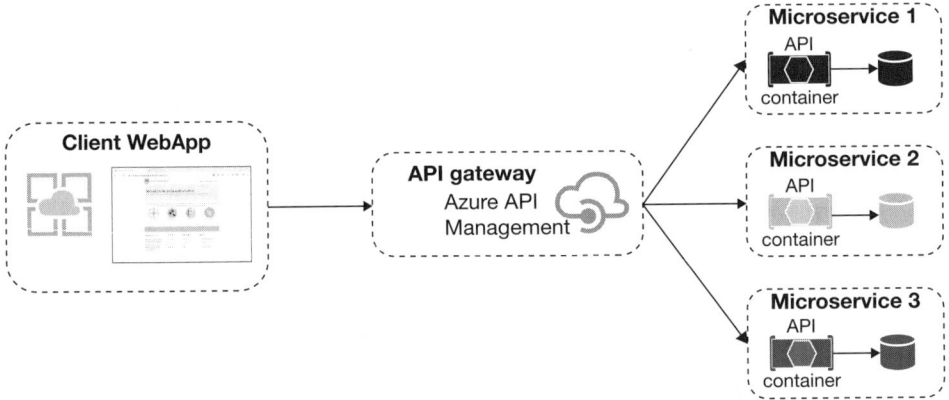

FIGURE 8-1 Communication across web applications to microservices through an API gateway where Azure API Management is used as an API gateway technology

The API gateway is the key component and cannot be overlooked when choosing the components and building the architecture for enterprise solutions that are based on the microservices architecture.

Azure API Management

Azure APIM (API Management) is a managed service on Azure that provides a very intuitive platform to publish, secure, transform, monitor, and maintain your APIs. Regardless of whether your service is a RESTful API, a SOAP-based API, or it's based on managed serverless services such as Azure Functions or Logic Apps, it can easily be configured in the Azure APIM and help you define policies and expressions to implement cross-cutting concerns.

Azure APIM provides a single front door for all the microservices running on containers, hosted on-premises, or anywhere else. The client application communicates to the endpoint of API Management, which then is routed to the back-end service, passing through different policies configured for that API.

Key benefits of using Azure API Management

Azure API Management comes with many benefits:

- Set different kinds of policies at inbound, back-end, outbound, and on-error levels.
- Configure a security server and use modern authentication protocols to validate and pass security tokens to back-end APIs.
- Developers can use the access portal and manage keys, documentation, and SDKs.
- Provides detailed reporting and analytics.
- Provides a RESTful API to manage Azure API Management. You can add APIs, invite users, manage products, and so on.

Set up Azure API Management

Azure APIM can be set up from the portal or by running ARM (Azure Resource Manager) or using a Terraform script. We will use Terraform to create the Azure APIM resource. However, for configuring APIs and Policies, we will explore the experience from the Azure portal side.

Let's start by creating a new Terraform file and naming it APIM.tf. Add it to the same folder where you have kept the Terraform files for other services. The following script provisions a basic APIM instance on Azure:

```
resource "azurerm_api_management" "apim" {
  name                = "oas"
  location            = "westeurope"
  resource_group_name = "OSS"
  publisher_name      = "OAS"
  publisher_email     = "info@oas.com"
  sku_name = "Developer_1"
}
```

We started by specifying the name and location. As per the canonical practice, it is a good approach to keep all the resources within one location to minimize latency issues. The resource group will be OSS in our case and the publisher_name and publisher_email can be specified as required. For the SKU_name, we have used Developer_1.

The above script can be executed using Terraform commands such as `Init`, `Plan`, and `Apply` in the sequence. To learn more about executing these commands, see the previous chapters. Once the script is successfully executed, the APIM will be provisioned, and you can verify it from the Azure portal.

Configure APIs in Azure APIM

To configure APIs in Azure APIM, there are multiple options:

- **Add an API manually**   With this option, you can start with creating a blank API template and manually define the API methods from the interface.

- **Import an Open API specification**   With this option, if your API supports an open API specification, you can pull the API methods by providing the complete URI of the API.

- **Import a SOAP API**   All SOAP-based services can be imported by providing a complete WSDL URL.

- **Import a SOAP API and convert to REST**   Importing an existing SOAP service can be done by referring to the WSDL URL, which then converts the SOAP service to a RESTful interface. To access this API, you can make REST calls to access it.

- **Import an API App**   If your API is hosted as an Azure App Service, you can also import it to APIM by selecting the API App option and choosing your API App.

- **Import a Function App**   If you have serverless apps running as Function Apps, they can also be imported to APIM by selecting the Function App option and choosing your Function App.

- **Import a Logic App**   If you have logic apps running, they can also be imported to APIM by selecting the Logic App option and choosing your Logic App.

- **Import a Service Fabric App**   If you have service fabric apps running, they can be imported, too.

- **Publish APIs directly**   Azure APIM also provides a RESTful endpoint to publish APIs directly by making an HTTP call to that endpoint.

> **NEED MORE INFO?**   To learn more about importing APIs for each option, see
> *https://docs.microsoft.com/en-us/azure/api-management/add-api-manually.*

To configure the OAS microservices, we will use the **Blank API** option and configure the APIs manually:

1. You need to first log in to the Azure portal and select the **APIs** option from the left pane, as shown in Figure 8-2.

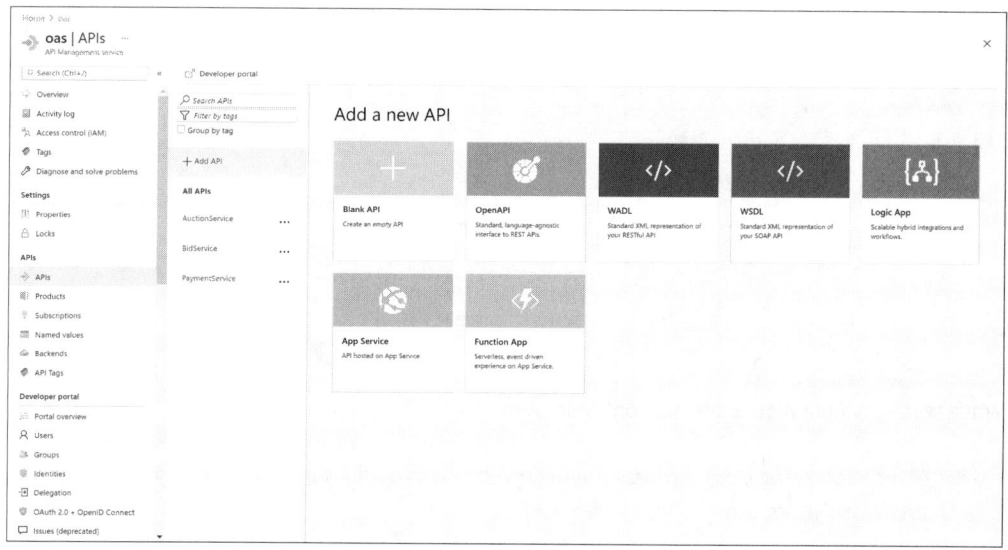

FIGURE 8-2 Add A New API

2. On the right side, click **+ Add API** to add a new API. You can choose the type of API you want to create by clicking the respective tile: Blank API, OpenAPI, WADL, WSDL, Logic App, App Service, or Function App. Select the **Blank API** option to start adding the API details for the auction service, as shown in Figure 8-3.

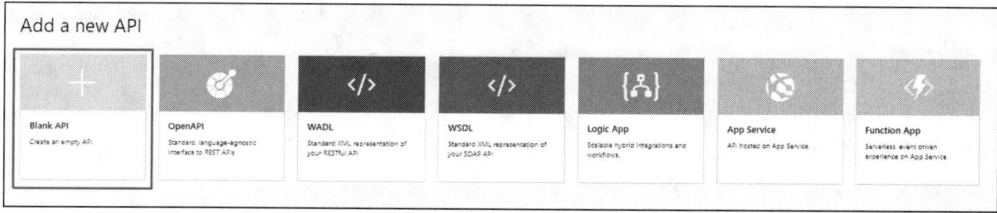

FIGURE 8-3 The options to configure and import APIs from various sources

3. On the **Blank API** page, enter the **Display name** as AuctionService, set the **Name** as auctionservice, set the Azure Kubernetes Service external IP address to the **Web service URL**, and set the API URL suffix to auctionservice, as shown in Figure 8-4.

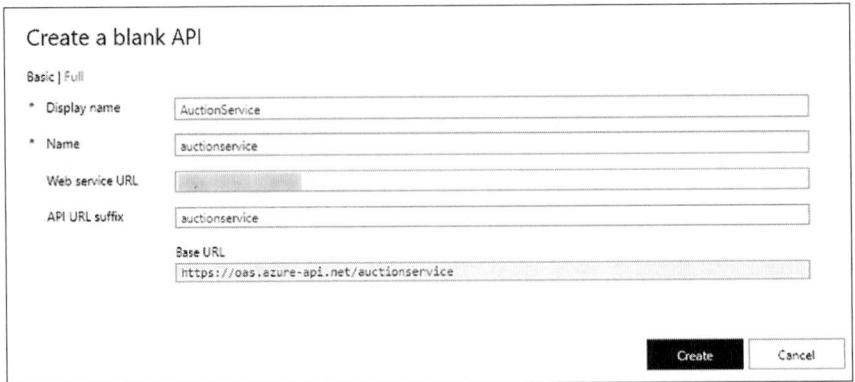

FIGURE 8-4 Configuration of the AuctionService API

4. After adding the details, press the **Create** button to add it into the **All APIs** list, which is shown in Figure 8-5.

FIGURE 8-5 The AuctionService API

5. So far, the basic API is configured, but we have not defined any web methods. To add them, select the AuctionService from the **All APIs** pane, as shown in Figure 8-5, and click the **Add operation** option, as shown in Figure 8-6.

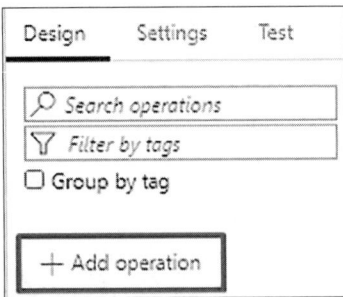

FIGURE 8-6 Add operation

6. Clicking **Add operation** opens a new blade on the right side of the screen where you can add web methods for the service. The first method we add will create auctions. Figure 8-7 shows the fields that must be configured.

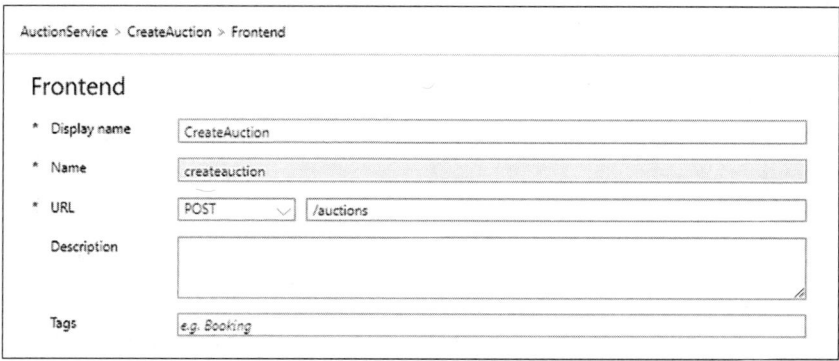

FIGURE 8-7 Adding an operation to create an auction

7. Specify the **Display name** as CreateAuction and the **Name** as createauction. Because you are creating a new auction, from the **URL** drop-down menu, select **POST**. In the field to the right of the URL drop-down menu, enter /auctions, which is the actual URL that will trigger the POST method.

8. Next, as shown in Figure 8-8, add another method to list all the auctions based on the auction ID. For this example, we will use the **auctionById** method and pass the parameter in curly brackets as shown below:

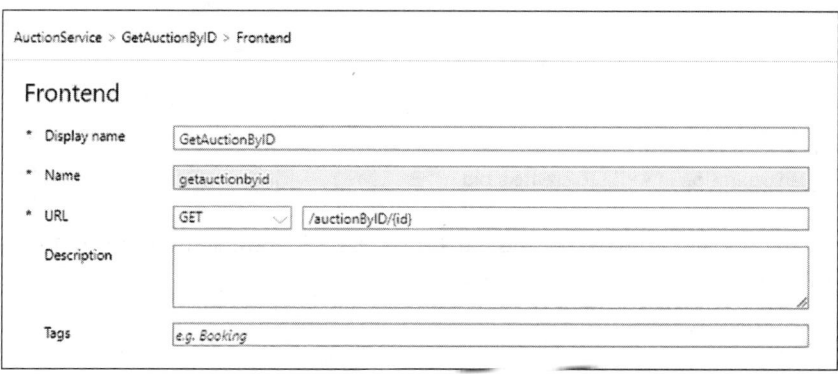

FIGURE 8-8 Adding an operation in APIM to get an auction based on the ID

Specifying the parameters inside curly brackets creates a template parameter that you need to pass when you make a call from the client application. To update the auction information when the bid is made, we will add another method, as shown in Figure 8-9.

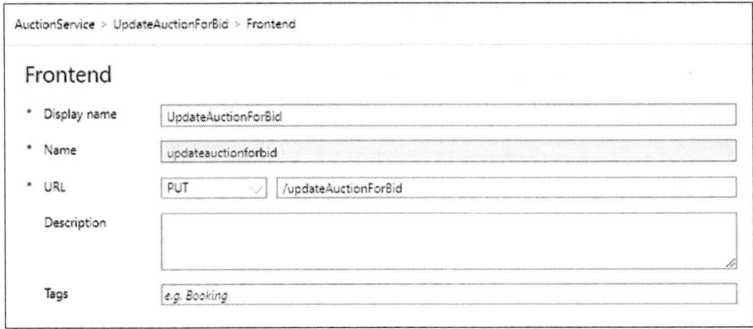

FIGURE 8-9 Adding an operation in APIM to update an auction

9. Because the request is to update the auction table, you should use a PUT request. Add other methods that you want to expose for the auction service and complete the configuration.

10. For the bid service, we will go with the same approach of adding a bid service using the Blank API option, as previously shown in Figure 8-3, and we will add CreateBid and GetAllBids methods. Figure 8-10 shows the CreateBid method.

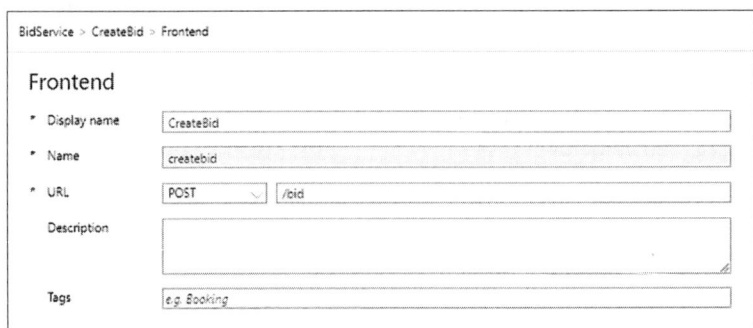

FIGURE 8-10 Adding an operation in APIM to create a bid

11. Because the method URL is the same, and the HTTP verb is different, it makes the HTTP GET request to the bid service. See Figure 8-11.

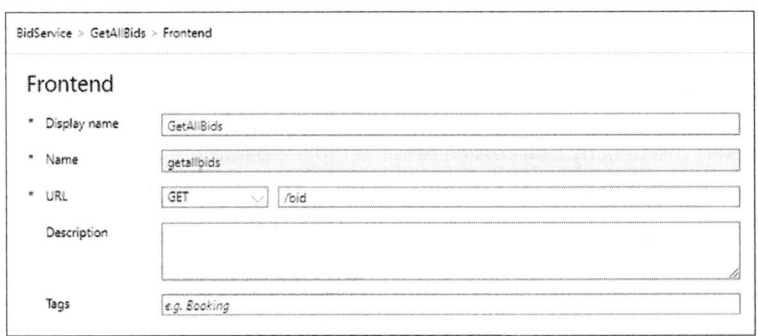

FIGURE 8-11 Adding an operation in APIM to get all bids

12. Lastly, for the payment service, we can use the same **Blank API** option as shown in Figure 8-3 and add operations to create and read auction payments. Figure 8-12 shows the creation of an auction payment.

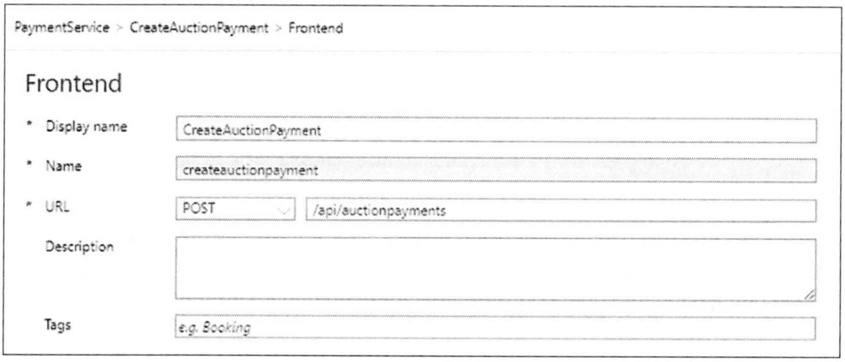

FIGURE 8-12 Adding an operation in APIM to create an auction payment

13. The HTTP verb will be POST, and the URL is /api/auctionpayments because the routing path defined for this service starts with /api. To read the auction payment, see the configuration for the GetAuctionPayment method in Figure 8-13.

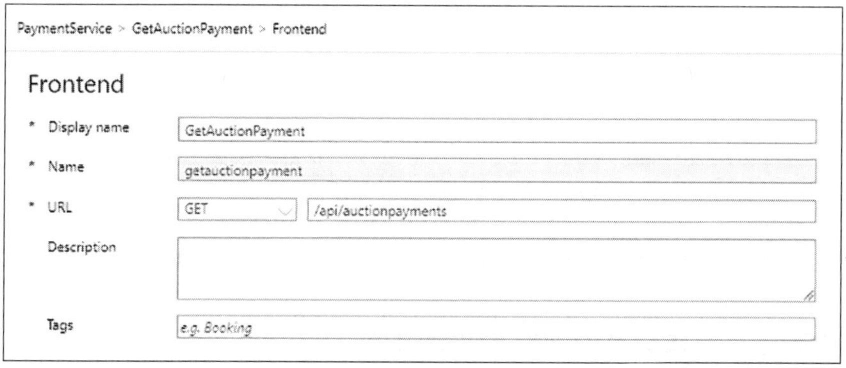

FIGURE 8-13 Adding an operation in APIM to get an auction payment

APIM provides an easy interface to quickly configure and import APIs using heterogenous standards and protocols. To test these APIs, we can use the **Test** pane from the APIM blade itself to make calls, or you can even test it from the browser or third-party tools such as Postman or Fiddler.

In APIM, we can enable or disable subscriptions to our APIs. If the subscription is enabled, we need to pass the subscription key for every request so the API can be accessed. However, if the subscription is disabled, users don't need to pass the subscription key, and the API can be accessed by anyone. To enable/disable subscription, you can select the API and go to the **Settings** pane and select or deselect the **Subscription Required** option in the **Subscription** section, as shown in Figure 8-14. If the **Subscription Required** option is not selected, no subscription key is required while making a call to that service.

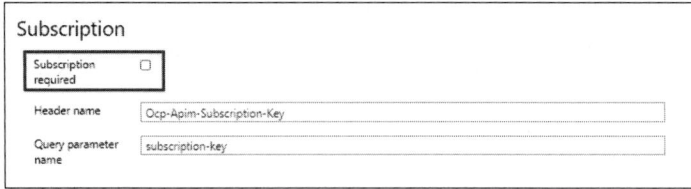

FIGURE 8-14 Enabling/disabling a subscription

You can pass the subscription key either in a request header or as a query parameter.

> **NOTE** The default **Header Name** is Ocp-Apim-Subscription-Key, whereas the default **Query Parameter Name** is subscription-key. However, this can be changed based on your own naming practice.

To get the subscription key, go to the **Test** pane by selecting the specific service from the **APIs** tab in APIM and look at the **HTTP Request** section, as shown in Figure 8-15.

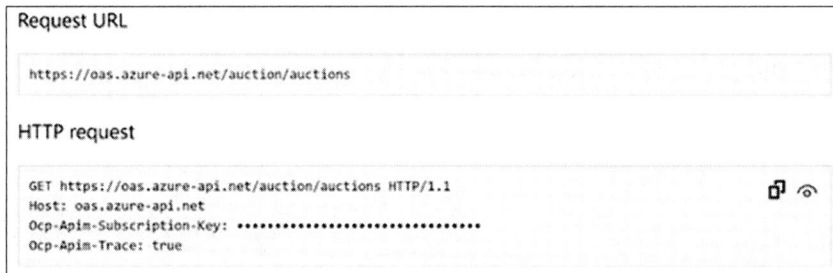

FIGURE 8-15 The subscription key can be copied to make calls to the APIs.

The subscription key is hidden by default, but you can click the eye icon to the right to view it.

Working with policies and expressions

The essence of the gateway is to control and secure your APIs by adding various rules and conditions. By default, the APIM provides policies that can be set at different levels to protect, control, and harden the security for your APIs. The policies can be set at the following stages:

- Inbound
- Backend
- Outbound
- On Error

Figure 8-16 illustrates the use of policies at each stage:

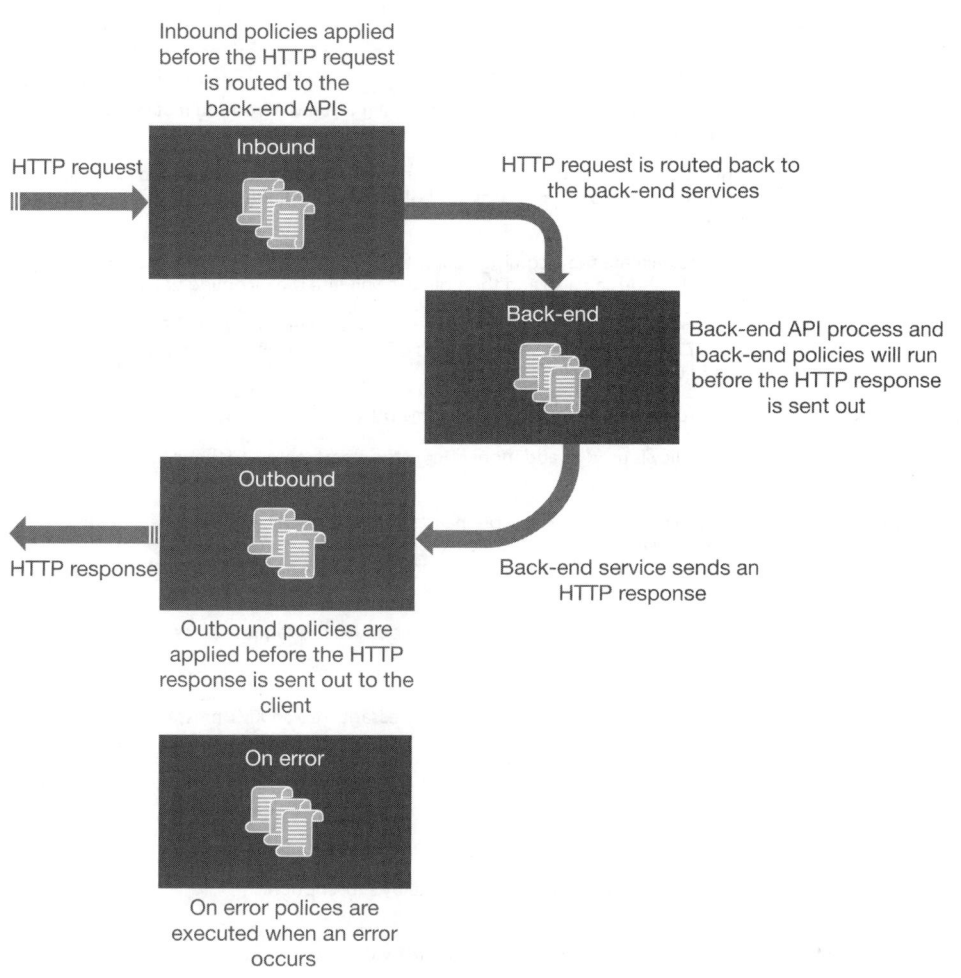

Inbound policies applied before the HTTP request is routed to the back-end APIs

HTTP request

Inbound

HTTP request is routed back to the back-end services

Back-end

Back-end API process and back-end policies will run before the HTTP response is sent out

Outbound

HTTP response

Outbound policies are applied before the HTTP response is sent out to the client

Back-end service sends an HTTP response

On error

On error polices are executed when an error occurs

FIGURE 8-16 Policies in APIM

Figure 8-16 depicts the three places where policies can be implemented in APIM and be executed sequentially during request and response to change an API's behavior. When the request comes to the APIM, the inbound policies are executed. Based on the policy configuration, the request can be manipulated, validated, and routed to the back-end APIs.

The back-end API policies are executed when the APIM calls the back-end APIs, and the outbound policies are executed when the APIM returns the response from the back-end APIs to the caller. If an error occurs at any scope, the on-error policies are executed. APIM provides a variety of policies that can be configured in specific areas.

You can also customize the behavior by using policy expressions. Some of the most common polices are shown in Table 8-1.

TABLE 8-1 Common Policies in APIM

Policy	Description
Mock response	Used to return mock responses to the caller instead of routing that request to the back-end API.
CORS	Add CORS (Cross Origin Resource Sharing) support to your back-end APIs. Instead of enabling CORS on your back-end APIs, you can enable it through this policy at the APIM level.
Validate JWT token	Used to validate the incoming bearer token from the caller. Instead of protecting your APIs, you can also add this policy to validate the incoming token.
Set request method	This policy is used to change the request method of the request. For example, if the HTTP request initiated by caller is GET and you want to change it to POST you would use this policy.
Rewrite URL	Used to convert a request URL to some other URL.
Set query string parameter	This policy is used to add more query string parameters to the request.
Set HTTP header	This policy is used to add more headers in the request. For example, if we need to add the Accept header to tell the back-end API to return a response in XML format, it can be done by adding the Accept header with a value such as application/xml.
Set body	This policy is used to add the message body in the request or response.
Find and replace in body	This policy is used to find and replace strings in the request or response body.
Convert JSON to XML	This policy is used to convert a JSON message into an XML message for both request and response.
Convert XML to JSON	This policy is used to convert an XML message into a JSON message for both the request and the response.
Restrict caller IPs	This policy is used to restrict the IPs of the caller.
Limit call rate by subscription	This policy is used to limit the call rate by subscription.
Limit call rate by key	This policy is used to limit the call rate by the key.

The policy definition is based on the XML format and is easily configurable at the product, API, or API operations levels. To see how the policy can be configured, let's configure the Validate JWT Token policy to validate the incoming token for the auction service.

To configure the Validate JWT Token policy, we need to first log in to the Azure portal and access the APIM instance. From the **All APIs** section, select **AuctionService** and then select **All Operations**, as shown in Figure 8-17.

FIGURE 8-17 Selecting All Operations

Select the **Inbound Processing** section and click **+Add Policy**. From the **Add Policy** page, select **Validate JWT**, as shown in Figure 8-18.

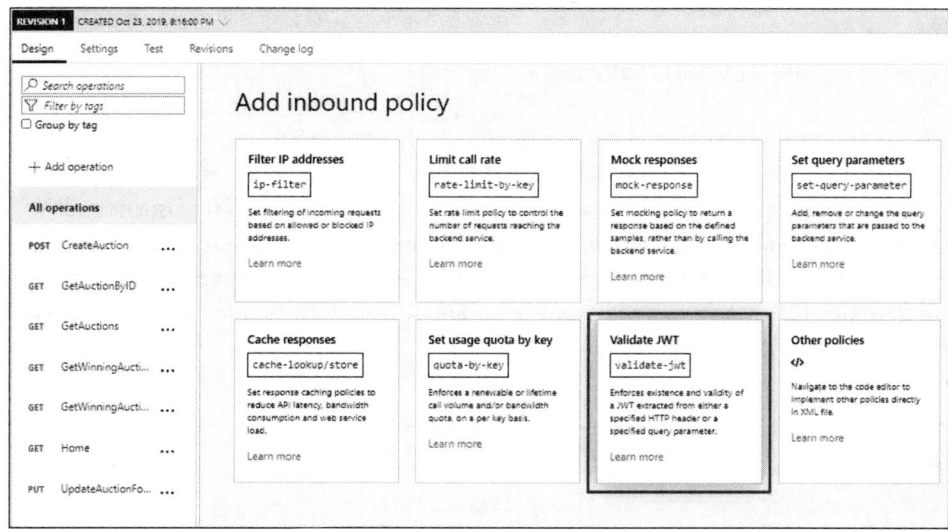

FIGURE 8-18 Policies available in APIM

Once you select the Validate JWT policy, it will show a window to configure the policy as shown in Figure 8-19.

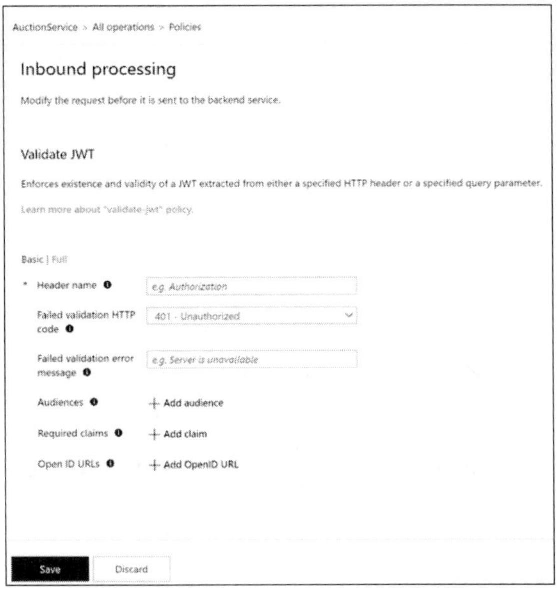

FIGURE 8-19   Configuring the Validate JWT policy in APIM

The **Header name** is the name of the request header holding the token. In this case, it will be Authorization. The **Failed validation HTTP code** is the response status our code will return if the token validation fails. The **Failed validation error message** is the user-friendly message that you can pass to the user when the validation fails. The **Audiences** field contains the application **Client ID** that is registered in Azure AD B2C. Finally, the **Open ID URLs** field contains the Open ID Connect metadata URL endpoint. After specifying the values, click **Save**. The policy is shown in Figure 8-20.

FIGURE 8-20   The Validate JWT Policy snippet

Strategies when using Azure API Management with Azure Kubernetes Services

For every application, there are different strategies and practices when configuring APIM. Provisioning APIM closer to your services reduces latency issues. If the services are hosted in the cloud, the APIM should be deployed canonically at the same location. However, sometimes we provision APIM in the cloud and configure VNET (Virtual Network) to allow access to on-premises resources, such as databases, message brokers, cache systems, and so on. In Kubernetes, we can consume our services running inside pods using Kubernetes services.

The following sections outline the types of services that allow an ingress of Internet traffic to reach the pods hosted inside the Kubernetes cluster (K8s).

K8s service with NodePort

The NodePort configuration allows you to access the pods inside the node using the node IP address. With NodePort, a port is assigned to all the nodes in your K8s cluster, and any request that is sent to that node IP is forwarded to the service. Figure 8-21 shows the architecture when using a K8s service with a NodePort configuration.

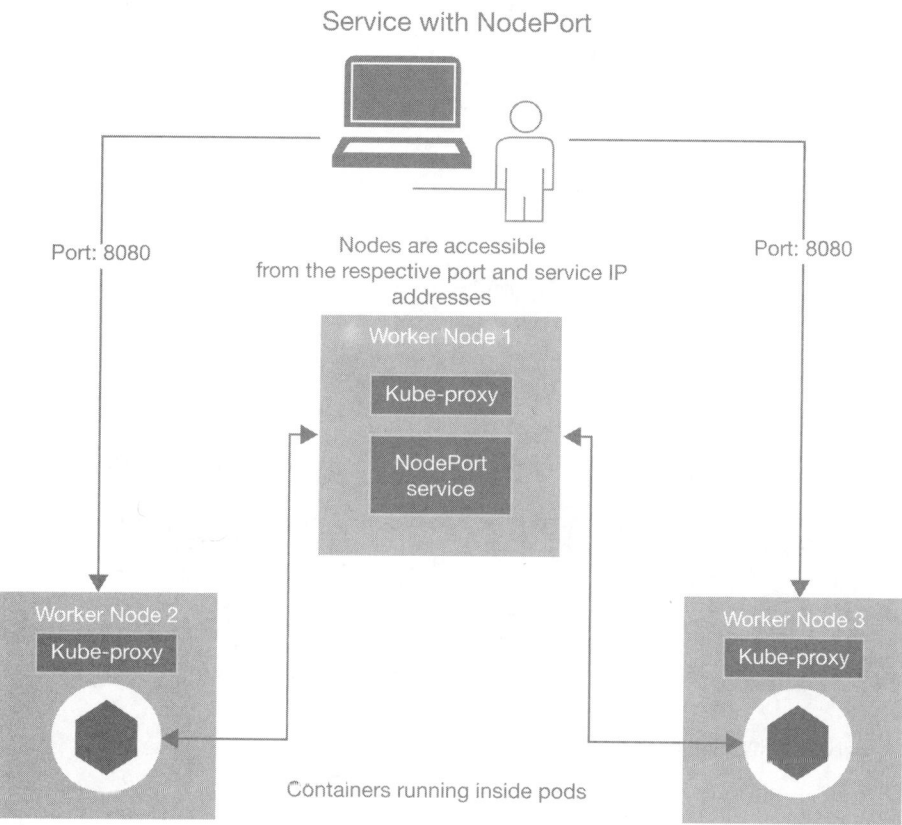

FIGURE 8-21 Service with NodePort configuration

In the NodePort architecture, the services can only be consumed within the K8s cluster, and the cluster nodes are provisioned with VNET private IP addresses.

K8s Service with the LoadBalancer configuration

The LoadBalancer configuration is used to create a LB (Load Balancer) that is exposed over a public IP address, and it allows an ingress of traffic to reach to the respective pods. Once the request comes to the load balancer, it is redirected to the respective pods associated with that service. Compared to the NodePort configuration, this is an internal LB available in K8s. With NodePort, you must rely on an external LB. Figure 8-22 shows the architecture for using a K8s service with a LoadBalancer configuration.

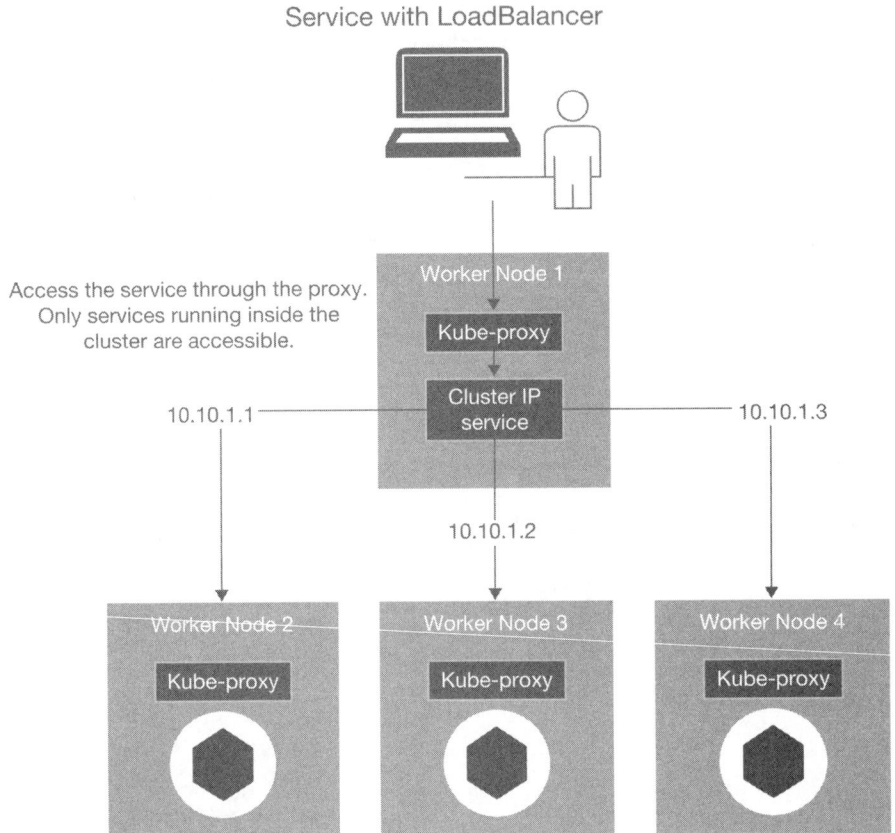

FIGURE 8-22 Service with LoadBalancer configuration

K8s service with a ClusterIP configuration

The ClusterIP type service is used to access the pod within a cluster. It cannot be accessed outside the cluster network. Figure 8-23 shows the architecture when using a K8s service with a ClusterIP configuration.

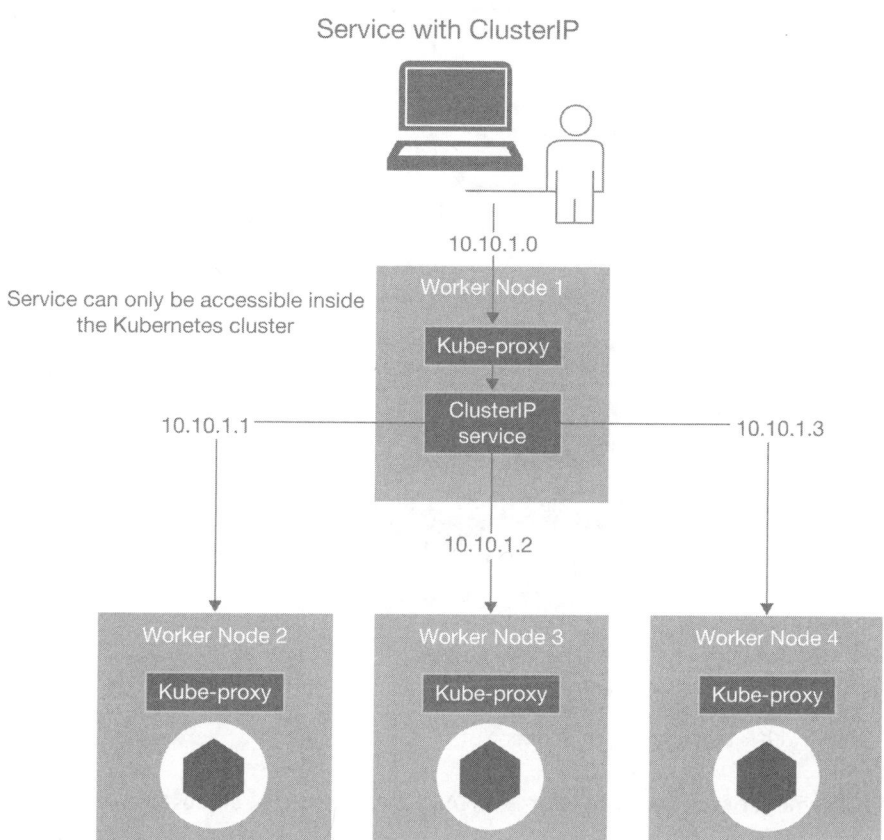

Service with ClusterIP

Service can only be accessible inside
the Kubernetes cluster

Containers running inside pods

FIGURE 8-23   Service with ClusterIP configuration

The ClusterIP configuration is normally used with applications that must be accessed internally from within the organizational network.

K8s service with the ExternalName configuration

The ExternalName type of service is used to map the request to an external service that is not hosted inside the K8s cluster. This can be any service that is accessible over a public endpoint. Figure 8-24 shows the architecture of using a K8s service with ExternalName configuration.

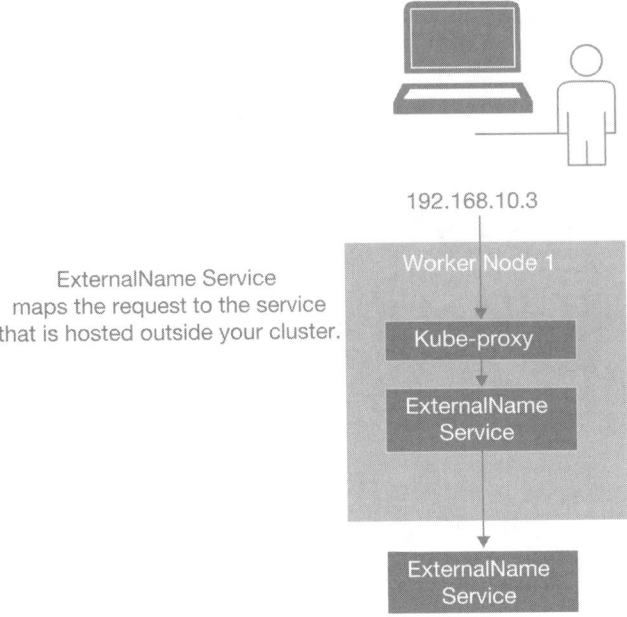

Service with ExternalName

192.168.10.3

Worker Node 1

ExternalName Service
maps the request to the service
that is hosted outside your cluster.

Kube-proxy

ExternalName
Service

ExternalName
Service

FIGURE 8-24   Service with ExternalName configuration

The ExternalName service creates an internal CNAME DNS entry that aliases to an external service.

Strategies to configure APIM with Azure Kubernetes Services

After going through a complete overview about the different kind of services that can be provisioned in Kubernetes, let's discuss some strategies about how the APIM can be configured with the services hosted inside AKS (Azure Kubernetes Services). There are three types of services that can be configured:

- Internet-accessible services in AKS
- Intranet-accessible services in AKS
- Ingress controller–accessible services in AKS

Internet-accessible services in AKS

In the AKS, to access the deployed and running containers inside pods, we need to configure Kubernetes service objects. The request can be made to the service, which routes traffic to the respective pods with which the service is associated. Figure 8-25 shows the use of APIM with services exposed publicly inside an AKS cluster.

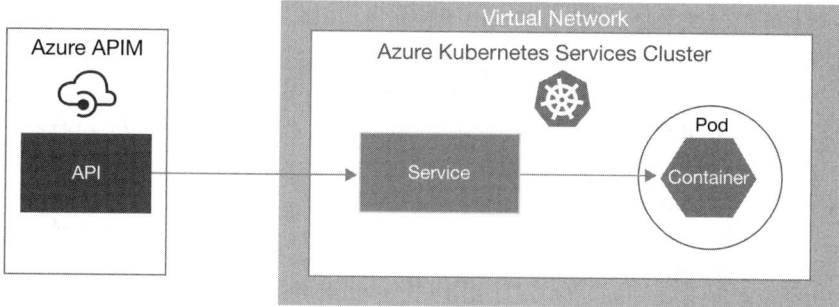

FIGURE 8-25 Architecture of services hosted in AKS and that are accessible over a public endpoint

This is easy to set up where the Azure API Management exposes APIs over the Internet and makes calls to the back-end services hosted inside the AKS cluster. The connection from APIM to the services hosted inside the AKS cluster is also exposed over a public endpoint. In this case, the service type can be a NodePort or a LoadBalancer. Each service hosted inside an AKS cluster exposes a public IP and on the APIM, it uses that public IP to configure it as a back-end service. This is the same strategy we followed when building the OAS.

Intranet-accessible services in AKS

Sometimes, the services do not need to be exposed over a public network, but the services must be accessible within an organizational intranet. In such cases, we need to deploy Azure APIM inside a virtual network. Figure 8-26 illustrates the deployment of APIM inside a VNET.

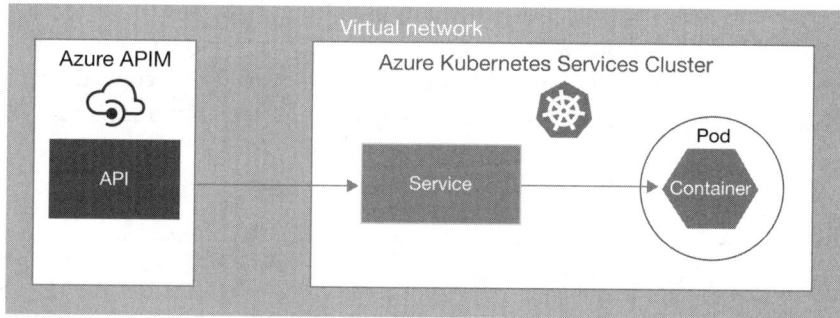

FIGURE 8-26 Running APIM and AKS in the same VNET

With this configuration, the APIM is deployed inside the cluster VNET and is accessible over the Internet. All requests that reach the APIM will be routed to the service hosted inside the cluster and will not be exposed over a public endpoint. In such cases, you don't need to implement security for your service or configure the Validate JWT policy on the APIM itself to validate the incoming token.

Ingress controller–accessible services in AKS

For security reasons, it is good practice to secure your APIs because they are publicly exposed. The APIM can be configured to validate the incoming access token and forward it to the respective services. However, if the scale of the application is large, then the validation of token needs an implementation for each service. To avoid this, we can also use another way of adding an ingress controller and delegate the responsibility of validating the incoming access token and then routing it to the back-end service. Figure 8-27 shows the use of an ingress controller inside the AKS cluster with APIM.

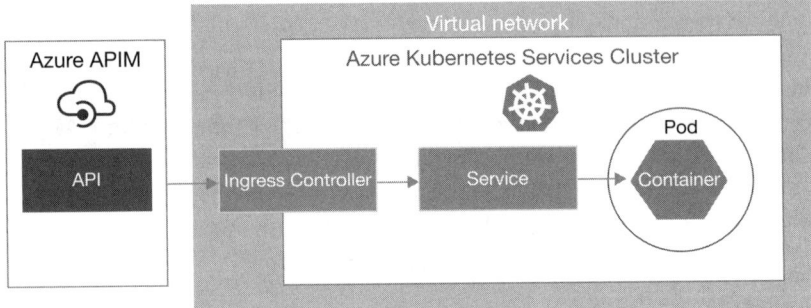

FIGURE 8-27   Adding an ingress controller that takes the traffic from APIM and then routes it to the service that is exposed within a cluster endpoint in AKS

With the architecture shown in Figure 8-27, the service can be accessed with a single base endpoint for the ingress controller. The traffic from the APIM goes to the ingress controller and is then routed back to the back-end services. The security can also be hardened on the ingress controller side, and the token validation will be done when the traffic reaches to the ingress controller. The services IP address will not be publicly exposed, which reduces security risks. When configuring services in AKS, ClusterIP is the service type supported for this scenario.

Summary

In this chapter, we learned the following topics:

- We began with a brief discussion about why the API gateway plays an important role in microservices architecture.
- We explored Azure API Management and its benefits.
- We set up Azure API Management and APIs.
- We discussed the policies and expressions in Azure API Management.

- We discussed deployment scenarios for services in Kubernetes cluster.
- We discussed a few strategies in terms of using Azure API Management with Azure Kubernetes Services.

In the next chapter, we will learn the importance of automation to build and deploy services, and we will explore how to set up the IaC (Infrastructure as Code) using Terraform. Also, we will set up continuous integration and continuous deployment to build and deploy services in Azure Kubernetes Services.

Chapter 9

Build and deploy microservices

In this chapter, you will:

- Summarize the importance of automation within builds and deployments of microservices through continuous integration and deployments
- Explore how to automate infrastructure provisioning using IaC with Terraform and demonstrate how this is done for the OAS
- Understand how to leverage Azure DevOps to build the pipelines needed for iterative improvements to our application
- Explore how to implement CI/CD using Azure DevOps

In Chapter 1, we emphasized many of the benefits of microservices. One of these benefits was the ability to generate an architecture that could maximize developer productivity by building out individual functional components. However, the catch was to figure out how to put these individual services together and push them out to create a single, unified application. In this chapter, we demonstrate how to build and deploy microservices to accomplish this. We will begin by talking about an overview on CI/CD and automation before going deeper in the tools used to build out our OAS and how to leverage these lessons to create pipelines yourself.

Continuous integration and continuous deployment

Let's think about the process of application deployment generically before we add specifics. When you are creating any application, regardless of architecture, the best practice is to first create a build and then release a build.

A build focuses on ensuring that compilation takes place, and all prerequisite checks are in place. Although this can differ based on the application and the kinds of languages that are being used, there is usually some sort of translation to ensure that the application is packaged as some form of an executable and can run with the proper dependencies and code libraries. This is usually seen as the first check for an application before deployment. You should be able to create other checks that will ensure that your code is working properly before deployment. Thus, many automated unit tests can be seen during the later stages of building and deploying applications.

Deployment focuses on the execution of the code to a particular medium. We leverage the objects created through the build process to run the commands that put them together in a singular application.

These steps are the backbone for many application deployments. However, as we saw in Chapter 1, simply building and deploying in aggregate is not a winning strategy for a microservices architecture. The key here is to use repetitive automation, constantly building new versions of the edited codebase and deploying them with agility. These processes are known as continuous integration and continuous deployment, which are often abbreviated as CI/CD.

CI/CD works primarily through pipelines, and many tools are available to help achieve this, such as Jenkins, Maven, and so on. For this project, we have used Azure DevOps (shown in Figure 9-1), which is a product centered on DevOps teams oriented to the methodology of the same name. In the case of true DevOps teams, we have developers pushing new code and improvement with operations folks assisting with this pipeline infrastructure to build and deploy the code, maintain databases, run regression tests, and so on. The mobility and agility of these same teams apply to functional teams that are building individual microservices. These functional teams are building out microservices and pushing out code, and there needs to be a process in place to build it out and then automatically deploy the code.

Figure 9-1 is the main page for the OAS Terraform project. You can generally access Azure DevOps by going to *dev.azure.com* and proceeding through the registration pages to initiate a similar project. You can see some of the settings on the right side, such as inviting others to your project, project stats, and the breakdown of the members of your project. The left-side blade houses the primary functions where you can navigate to the summary or wiki of the project, repos that can house your project code, boards to run scrum meetings and leverage agile methodology, pipelines for building our continuous integration and deployment, and more.

FIGURE 9-1   An Azure DevOps project

Tools like Azure DevOps address CI/CD with the use of pipelines, which enable you to use a YAML file of coded instructions declaring how an application, infrastructure, or other asset should be deployed or built. If you are not comfortable with YAML, you even have the option to leverage the built-in UI in DevOps to write out these instructions easily. Later in this chapter, we will illustrate this UI. Before diving deeper into more complex examples, we want to discuss some capabilities of Azure DevOps by using Infrastructure as Code (IaC).

Automating infrastructure through Infrastructure as Code

In this section, we will illustrate how to use automation tools and continuous integration and deployment. We will be using Azure DevOps to demonstrate this, and we will set up pipelines to show how to enable Infrastructure as Code (IaC).

As we discussed in Chapter 4, "Develop microservices and front-end applications," IaC tools can help us automate the deployment of our application infrastructure using scripts. In Azure, we have tools known as Azure Resource Manager (ARM) templates that automate deployment with JSON and an interpretative engine to run and build the components.

In the open-source community, Hashicorp's Terraform is often used. Terraform is an IaC construct that can leverage different connectors to create infrastructure components in multiple clouds, such as Azure, AWS, and more. Its flexibility makes it attractive to many enterprise customers who are delving into polycloud/multicloud environments, which is why we have used it as our CI/CD example.

OAS Infrastructure as Code with Terraform

See Listing 9-1, which is a Terraform script that maps to the infrastructure components needed for our OAS application.

LISTING 9-1   Terraform script for infrastructure components

```
provider "azurerm" {
    features {}
  }


  resource "azurerm_resource_group" "rg" {
   name     = "OAS"
   location = "East US"
}


resource "azurerm_container_registry" "acr" {
    name                   = "oasregistry"
    resource_group_name    = azurerm_resource_group.rg.name
    location               = azurerm_resource_group.rg.location
    sku                    = "Standard"
```

```
      admin_enabled            = true
}

  resource "azurerm_resource_group" "aks" {
    name     = "oas-k8s"
    location = "East US"
  }

  resource "azurerm_kubernetes_cluster" "aks" {
    name                = "oask8scluster"
    location            = azurerm_resource_group.aks.location
    resource_group_name = azurerm_resource_group.aks.name
    dns_prefix          = "oas-k8s"

    default_node_pool {
      name       = "default"
      node_count = 1
      vm_size    = "Standard_DS2_v2"
    }

    identity {
      type = "SystemAssigned"
    }

    addon_profile {
      aci_connector_linux {
        enabled = false
      }

      azure_policy {
        enabled = false
      }

      http_application_routing {
        enabled = false
      }
```

```
      kube_dashboard {
        enabled = true
      }

      oms_agent {
        enabled = false
      }
    }
  }


  resource "azurerm_app_service_plan" "appserviceplan" {
  name                = "oas-appserviceplan"
  location            = "East US"
  resource_group_name = azurerm_resource_group.rg.name

  sku {
    tier = "Standard"
    size = "S1"
  }
}


resource "azurerm_app_service" "appservice" {
  name                = "oasapp"
  location            = azurerm_app_service_plan.appserviceplan.location
  resource_group_name = azurerm_app_service_plan.appserviceplan.resource_group_name
  app_service_plan_id = azurerm_app_service_plan.appserviceplan.id

}


resource "azurerm_cosmosdb_account" "cosmosdb-account" {
  name                = "cosmos-db-oas"
  location            = "East US"
  resource_group_name = azurerm_resource_group.rg.name
  offer_type          = "Standard"
  kind                = "MongoDB"
```

```
  consistency_policy {
    consistency_level        = "BoundedStaleness"
    max_interval_in_seconds = 10
    max_staleness_prefix     = 200
  }
  geo_location {
    location            = "eastus"
    failover_priority = 0
  }

}

resource "azurerm_cosmosdb_mongo_database" "mongodb" {
  name                    = "biddb"
  resource_group_name = azurerm_cosmosdb_account.cosmosdb-account.resource_group_name
  account_name            = azurerm_cosmosdb_account.cosmosdb-account.name
}

resource "azurerm_sql_server" "oasresource" {
  name                          = "oassqlsvr"
  resource_group_name           = azurerm_resource_group.rg.name
  location                      = "East US"
  version                       = "12.0"
  administrator_login           = "sqladmin"
  administrator_login_password = "P@ssw0rd!@#"

}

resource "azurerm_sql_database" "oasresource" {
  name                = "auctionpaymentdb"
  resource_group_name = azurerm_sql_server.oasresource.resource_group_name
  location            = azurerm_sql_server.oasresource.location
  server_name         = azurerm_sql_server.oasresource.name

}
```

```
resource "azurerm_mysql_server" "resourcevalues" {
  name                = "mysqloas"
  location            = "East US"
  resource_group_name = azurerm_resource_group.rg.name


  administrator_login          = "mysqladmin"
  administrator_login_password = "P@ssw0rd!@#"


  sku_name   = "B_Gen5_2"
  storage_mb = 5120
  version    = "5.7"


  auto_grow_enabled           = true
  backup_retention_days       = 7
  ssl_enforcement_enabled     = false
}
```

As you can see at the top of the code in Listing 9-1, we declared the use of the Azure RM provider to provision the related resources. Next, we built out different components with their respective specifications. We first declared an Azure resource group for the OAS and then built out the ACR using those particular names and locations. We used a similar patterning to build out a resource group for AKS, an AKS cluster, an App Service Plan, the App Service itself, a Cosmos DB account and the respective database, an SQL server and an SQL database, and a MySQL server.

> **MORE INFO**   If you have any questions on the syntax and how we were able to create this script, see the public documentation for Azure on Terraform at *https://docs.microsoft.com/en-us/azure/developer/terraform/.*

This script is named `main.tf` and is located in the Terraform folder in the `OAS-main-tf` branch. When you are scripting your components, make sure to similarly label that file as your `main.tf` file and keep it in a branch in the Terraform folder.

Build a pipeline or continuous integration

Once your repositories contain the `tf` files, you can start the process of building it with continuous integration. To do this, we will use the **Pipelines** feature in Azure DevOps located on the left pane menu. You will be prompted to identify where your code is located, as shown in Figure 9-2.

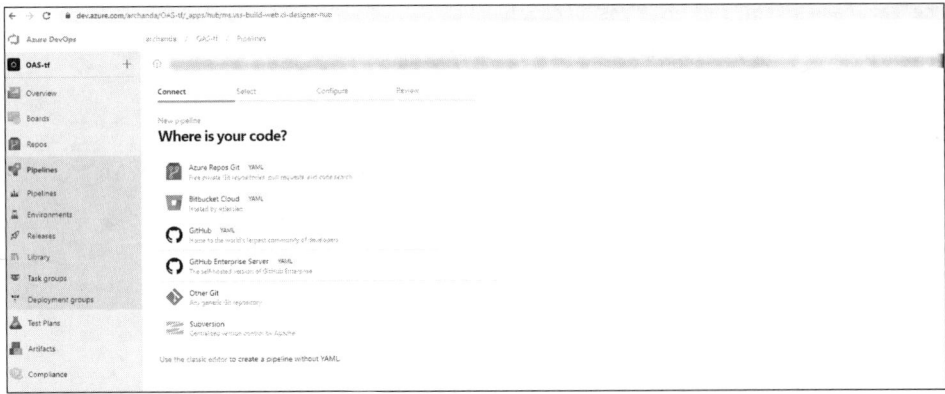

FIGURE 9-2 Creating a new pipeline

Then you select the option to connect your code base to the pipeline, so it knows where to draw the code from. If you continue using this process, you will be prompted to use YAML files. YAML files contain code to define the modern deployment experience. For large enterprise deployments, we will be using YAML to write out how we want to build out our components. Later in this chapter, we will show the YAML configurations for one of the microservices in the OAS later, but for the sake of this example, we will show the parts you are building in the classic editor UI. See Figure 9-3, which shows the screen after selecting the option to use the classic editor, it is shown Figure 9-2.

FIGURE 9-3 Classic Editor UI for Pipelines

This is the classic editor, and you can pick the code location and follow a deployment method without using YAML.

In Figure 9-3, you pick **Azure Repos Git** under **Select A Source** because this is where the main.tf file resides. Make the proper selections from the **Team Project**, **Repository**, and **Default Branch For Manual And Scheduled Builds** drop-down menus. For this example, we chose **OAS-tf**, **OAS-main-tf**, and **Master**, respectively.

After filling these out, you are asked to select a template of a job, as shown in Figure 9-4.

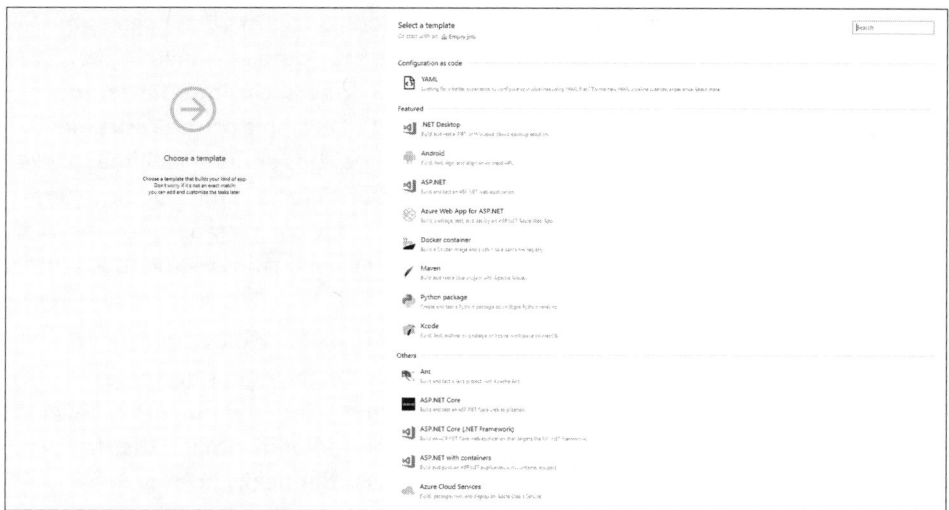

FIGURE 9-4 Select a template

Essentially, these templates are shortcuts that help you deploy certain kinds of commonly used projects faster. This can range from .NET to Android to Docker containers and many more. However, we will take on an empty job to illustrate a non-hydrated template, so we can walk step by step through the deployment.

> **TIP** Most enterprise deployments today use YAML, and it is highly recommended that you leverage its flexibility. We covered YAML in Chapter 5, "Microservices on containers."

Figure 9-5 shows a pipeline being built out in Azure DevOps. On the **Tasks** tab, click **Pipeline**, and at the right, enter the **Name** and choose the **Agent pool** and **Agent Specification** from the respective drop-down menus. In this example, **Azure Pipelines** has been selected from the **Agent pool** drop-down menu, and **vs2017-win2016** has been selected as the **Agent Specification**. These options define the context in which the commands you will execute will operate in, such as using the Windows operating system.

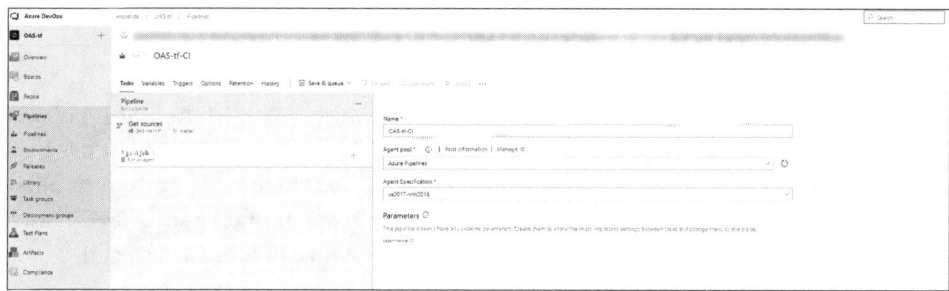

FIGURE 9-5 Pipeline build settings

Figure 9-5 shows the start of the pipeline without any additional tasks that need to be performed in the deployment. In essence, it is a blank slate and doesn't do anything at the moment. At the top is a row of tabs: **Tasks**, **Variables**, **Triggers**, **Options**, **Retention**, **History**, and **Save & queue**. We will primarily be focusing this example on the **Tasks** menu, but if you are interested in diving deeper into all the Azure DevOps options, feel free to take a look at the Azure DevOps documentation on Microsoft Docs. We have the workflow here for the pipeline, where we see the **Get sources** section is already completed from our earlier steps, and we now need to add new agent jobs, which are the commands we run to execute our deployment.

In Figure 9-6, we begin by adding in two agent jobs. We select the plus sign next to **Agent job 1** and search for and add **Copy Terraform Files**. On the right side, the pane shows the parameters for the **Copy Terraform Files** job. In the **Task version** drop-down menu, **2*** has been selected. **Copy Terraform Files** has been entered into the **Display Name** field, **Terraform** has been entered into the **Source Folder** field, and $(build.artifactstagingdirectory)/Terraform has been entered into the **Target Folder** field. These parameters again declare how to run the **Copy Terraform Files** job, so in this case, we are specifically copying all of the files in the folder **Terraform** and putting them in a staging area for execution.

> **MORE INFO**   To learn more about Azure DevOps, see
> *https://docs.microsoft.com/en-us/azure/devops/?view=azure-devops*.

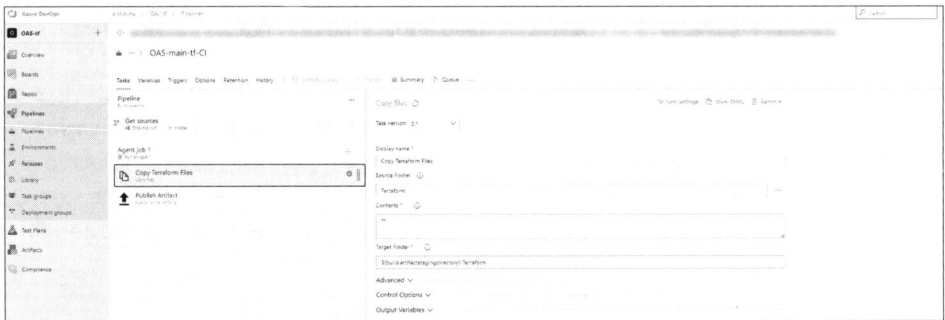

FIGURE 9-6   Agent job 1 for Build Pipeline

In Figure 9-7, we focus on the second task of **Agent job 1** which is **Publish Artifact**. On the right, the pane for this task is shown. In the **Task version** drop-down menu, **1.*** has been selected. In the **Display name** field, **Publish Artifact** has been entered; in the **Path to publish** field, $(Build.ArtifactStagingDirectory) has been entered; in the **Artifact name** field, **drop** has been entered; and in the **Artifact publish location** field, **Azure Pipelines** has been selected.

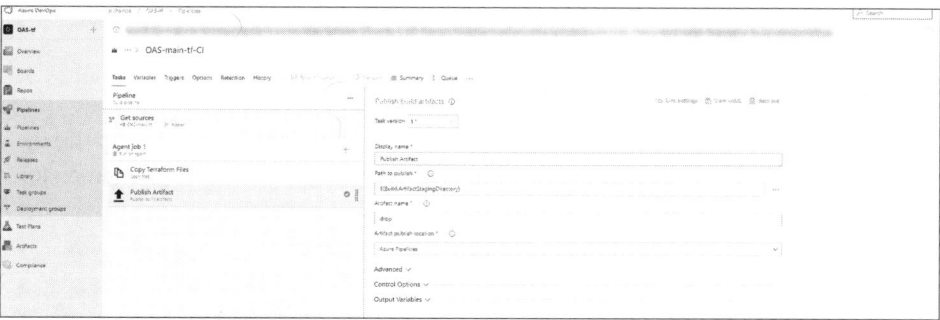

FIGURE 9-7 Publish Artifacts

In the operation shown in Figure 9-7, we are publishing build artifacts to use in our deployment pipelines. In order to use these files in the CD pipeline, we need to publish the artifacts we had copied over in the **Copy Terraform Files** task. Figure 9-7 shows the configuration for the task to publish the artifacts where we publish and drop the Terraform files into the result of this build pipeline so we can connect it to the deployment pipeline. That is the entire build for our infrastructure as code, so once these changes are complete, you can click **Save & queue** to automatically run the pipeline, as you can see in Figure 9-8. This figure shows the build being run and the actions that will be taken for the defined job operations. After all this is computed, the build artifact is created and can be used within the deployment.

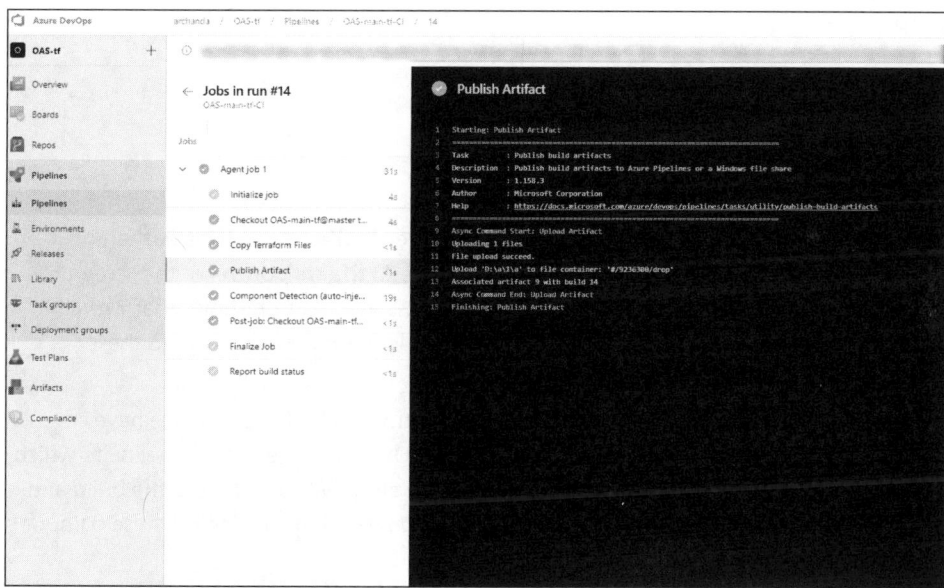

FIGURE 9-8 The build pipeline running

> **TIP** If you select each of the steps in a job, you can see the details for how each operation was carried out. If you have bugs in your build, you can come here to help troubleshoot the problem.

Now we can use the resulting artifact as part of our deployment pipeline.

Deployment pipeline or continuous deployment

Now, let's build our deployment pipeline, also known as a release pipeline. You can do this by referring back to Figure 9-2 and selecting the **Releases** option in the left pane. The screen for creating this pipeline will look like the one in Figure 9-9.

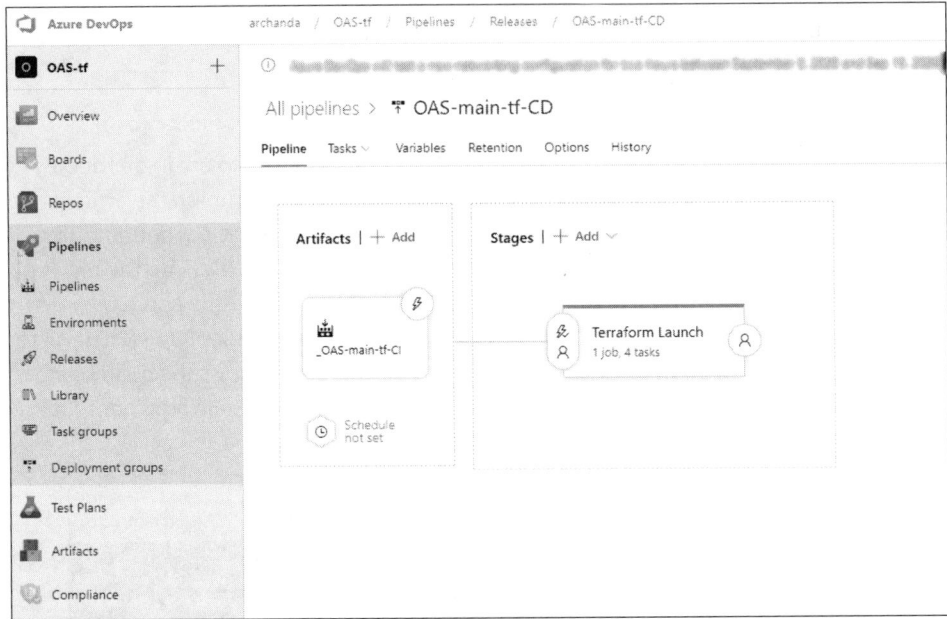

FIGURE 9-9   Creating a new release pipeline

Figure 9-9 shows the pipeline view of the release. Under **Artifacts**, we have gathered our resulting build that was the result of Figure 9-8. In the **Artifacts** portion of the screen, you can select that build. The job on the right side in the **Stages** section shows that we will perform a **Terraform Launch** command with four tasks defined. We will dive into these configurations next.

As you can see, we have configured our artifact to be the result of the build we have created. We simply select our build created from this Azure DevOps project, the pipeline to which it is related, the latest version of the build, and the source alias. We can now use the build and operate on it with a job that uses Terraform functions and tasks that will deploy these components in Azure.

Figure 9-10 shows the tasks in our Terraform release pipeline. The tasks are Install Terraform 0.1.2.3, `terraform init`, `terraform plan`, and `terraform apply`.

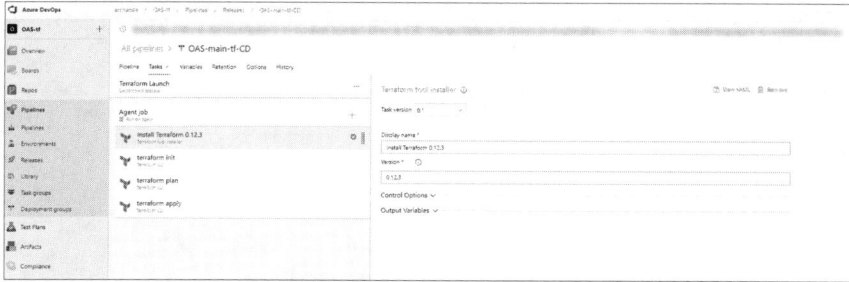

FIGURE 9-10 Tasks in the release pipeline

The tasks shown in Figure 9-10 and successive figures are broken down as follows:

1. First, we install Terraform so that we can run Terraform commands. We can see this in Figure 9-10 where we set the version of Terraform that we want to install.

2. The second task contains the Terraform commands: `terraform init`, `terraform plan`, and `terraform apply`. We configure the UI to use the drop folder that was created to store the artifacts for each task. Further, you have to configure these commands by selecting the Terraform CLI from the Marketplace, as you see in Figure 9-11, and then you change the commands in the configuration to fit your needs, as shown in Figure 9-12. The `terraform plan` and `terraform apply` commands have similar configurations to what you see in Figure 9-12.

FIGURE 9-11 Selecting the Terraform CLI from the Marketplace

FIGURE 9-12 Configuring the `terraform init` command

3. Next, we create the release, which should use the most recent version of your artifact so that the build is deployed using the commands we have defined here.

4. The deployment process is similar. The logs are shown in Figures 9-13 and 9-14.

FIGURE 9-13 High-level deployment process steps

FIGURE 9-14 The logs for the terraform apply task

5. As the deployment process is running, you will see the actions taking place in the Azure portal, where you see the components being created. Figure 9-15 shows the OAS resource group we deployed, which contains several of the components we just deployed.

FIGURE 9-15 OAS resource group containing deployed resources

Now that we have set up the CI/CD pipelines for our IaC, we can take this to the next level by setting up CI/CD for microservices using our OAS as an example.

Building CI/CD pipelines for the OAS microservices

In this section, we will showcase our build and deployment pipelines for the auction service because the pattern of build and deployment is similar to our other microservices. However, before we go deeper into the process of how to do this, we need to make a distinction that will lead to best practices.

CI/CD pattern and best practices

When we think about continuous integration, we are thinking about the build process (manual, automatic, and so on). Depending on the way your particular process is set up, differentiations can exist. That being said, one such practice is to create a trigger such that when a change is made to the code base, we build out a container image for the build. After this image is built, it is then pushed to the container registry. The deployment process will then use the image with kubectl commands to run a container application, which is one of our microservices.

We've seen many customers use this process, time and time again, so then, what's wrong with it? The problem resides in the overhead of many container images. Depending on the maturity of the enterprise, you can see many builds being kicked off within the span of a single week, which can potentially result in hundreds—maybe thousands—of builds being kicked off. Assuming weekly sprints, you must decide which of these builds will be deployed to production, which results in a disjointed process that moves away from what CI/CD is supposed to accomplish. Also, maintaining that many images in a container registry will cost you a lot. Do not make your registry a repository for images.

The alternative is to continue creating builds that are kicked off because of changes in the code base. Instead of pushing those images to the container registry, you instead want to save the container images as `.tar` files in the build server. This will depend on the type of tool you are using. For example, we used Azure DevOps, but you could use an on-premises option directly on your machine. By not storing your builds in the container registry, you are saving on costs and keeping the images in a little to no cost area. Now when you want to roll out a working build to production, you can push that build to ACR and then deploy the most recent version in a release pipeline.

Deployment patterns

Before we get into the actual pipelines, we should also call out some information on the different kinds of deployment patterns for microservices applications. For the OAS, we chose to use a service instance per container pattern, which means that each container holds a single microservice instance, as shown in Figure 9-16.

Service instance per container

FIGURE 9-16 Service instance per container deployment pattern

In Figure 9-16, there are two separate entities, or hosts (denoted by the two large rectangles), and each entity contains two virtual machines (denoted by the smaller squares with the computer symbol). In a single host, each of the virtual machines has a microservice (shopping cart and currency symbols) running in a container (hexagon) and is replicated to the other host as well.

Although this is the methodology we used for the OAS, other patterns are available, such as multiple service instances per host, service instance per host, and service instance per VM. Figure 9-17 shows the three VM deployment patterns: multiple service instances per host; service instance per host; and service instance per VM.

- In the multiple service instances per host pattern, there are three hosts and two microservices in each host.
- In the service instance per host pattern, there are three different hosts, each with a single microservice replicated across all three.
- In the service instance per VM pattern, there are two hosts, each running two VMs. In a single host, each of the VMs runs a different microservice, which is replicated to the other host.

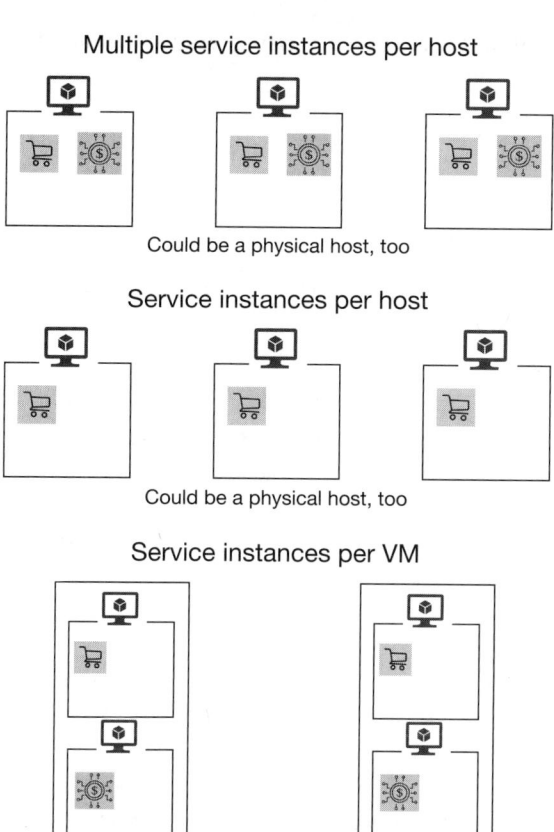

Multiple service instances per host

Could be a physical host, too

Service instances per host

Could be a physical host, too

Service instances per VM

FIGURE 9-17   VM deployment patterns

Your current resources can determine which pattern you choose to go with, but performance considerations led to us choosing the service instance per container pattern for the OAS.

The auction service build pipeline

Let's take a look at the pipeline for the auction service, which is shown in Figure 9-18.

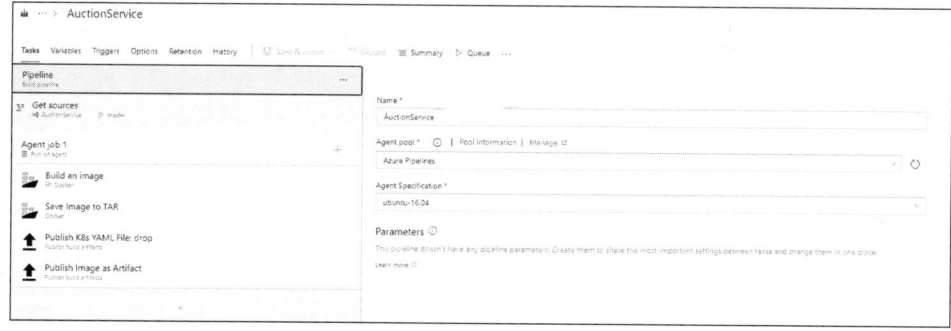

FIGURE 9-18   OAS auction service build pipeline in Azure DevOps

We are gathering the sources from the auction service repo that was created in Chapter 4, and we have set up an agent job to run several tasks:

- Build an image in Docker.
- Save the image as a TAR file.
- Publish a K8s YAML file.
- Publish the image as an artifact.

You will notice that this process mimics our guidance from earlier in the chapter, where we recommended creating TAR files that exist within a drop folder rather than pushing them to the container registry.

The configuration needed for these commands isn't too complex when compared to the Terraform configuration. When deploying the image, you want to make sure that you have selected the right registry and the proper Azure subscription from which you will perform the operation. You will be using the default Dockerfile for this service, which will be pulled from the repository. When you save an image to a TAR file, you want to use the Docker save task and use the following arguments:

```
--output $(Build.ArtifactStagingDirectory)/$(AuctionServiceImageName).image.tar
$(ContainerRegistryName)/$(AuctionServiceImageName):$(Build.BuildId)
```

This command adds a `BuildId` to the TAR file and then saves it in the corresponding folder.

Another task publishes the Kubernetes YAML file and adds it to the drop folder. As you know, the YAML file contains the deployment configuration as code and is the one we referred to in Chapter 5. We have placed this file in our repository because adding it to the build is vital when the build is being used for the deployment. The vital settings to input here are the path and where to publish (the drop folder).

Lastly, we need to publish the image as an artifact to the drop folder for use in the release pipeline.

We used these plug-and-play features here to leverage the power of multiple tools, such as Docker, Terraform, and even kubectl, as you will see shortly. Using a variety of features ensures that you can use different kinds of tasks, based on your needs.

If you have a scenario in which you need a specific task, you should leverage the Marketplace, which is shown in Figure 9-19. From the Marketplace, you can install various toolkits and extensions that allow you to use different tasks and variables for a variety of build and release pipelines.

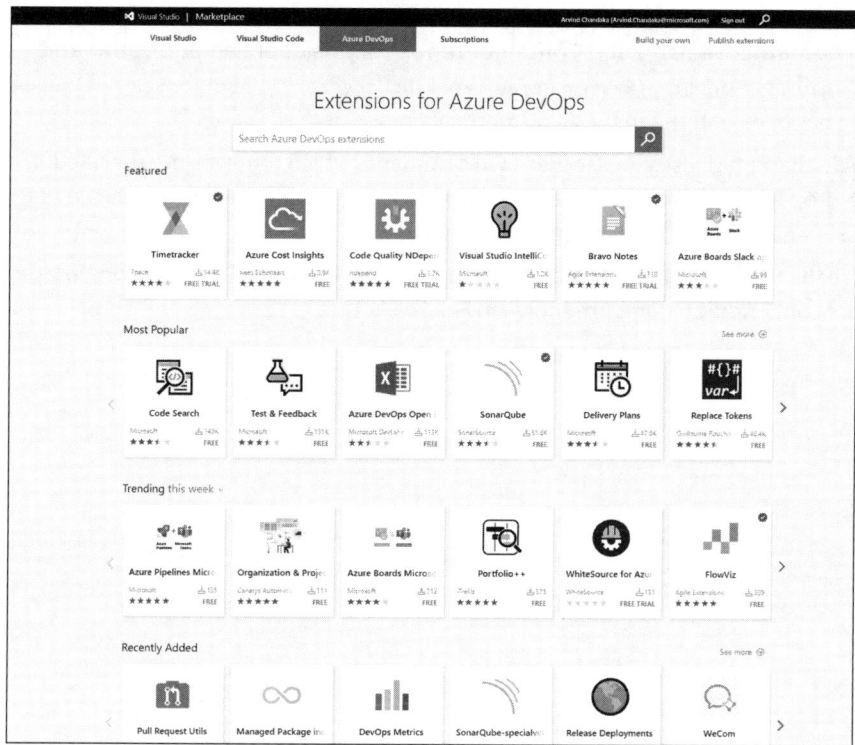

FIGURE 9-19 Azure DevOps Marketplace

Auction service deployment pipeline

Figure 9-20 shows the release pipeline for the auction service, where you can see four tasks that help deploy the final microservice.

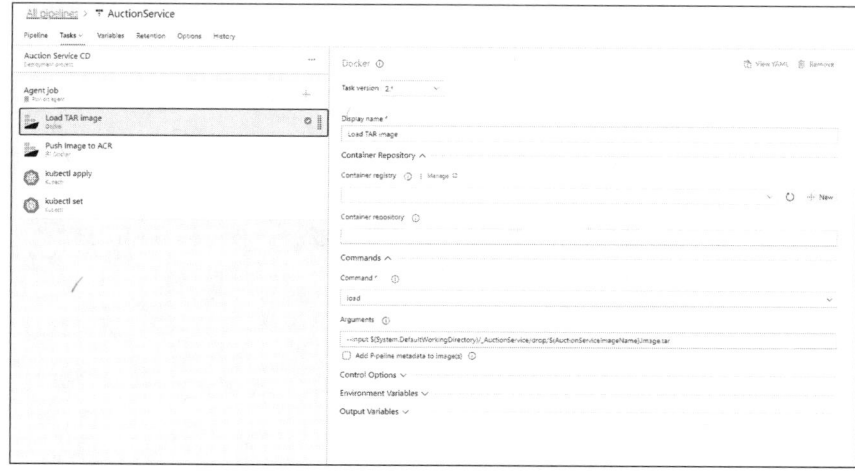

FIGURE 9-20 Auction service release pipeline

In Figure 9-20, the **Task version** is set to **2.***; **Load TAR Image** is entered into the **Display name** field; the **Container registry** and **Container repository** fields are empty; **Command** has been set to `load`, and `–input $(System.DefaultWorkingDirectory)/_AuctionServiceImage-Name).image.tar` has been entered in the **Arguments** field.

The **Load TAR image** task uses the Docker `load` command, which retrieves the specified image from the drop folder, setting it as the new deployment. This image is pushed to the ACR because it is the image that has been selected for deployment.

Next, we have the **Push Image To ACR** task. The **Container Registry Type**, **Action**, **Image Name**, and more configuration items are shown in Figure 9-21.

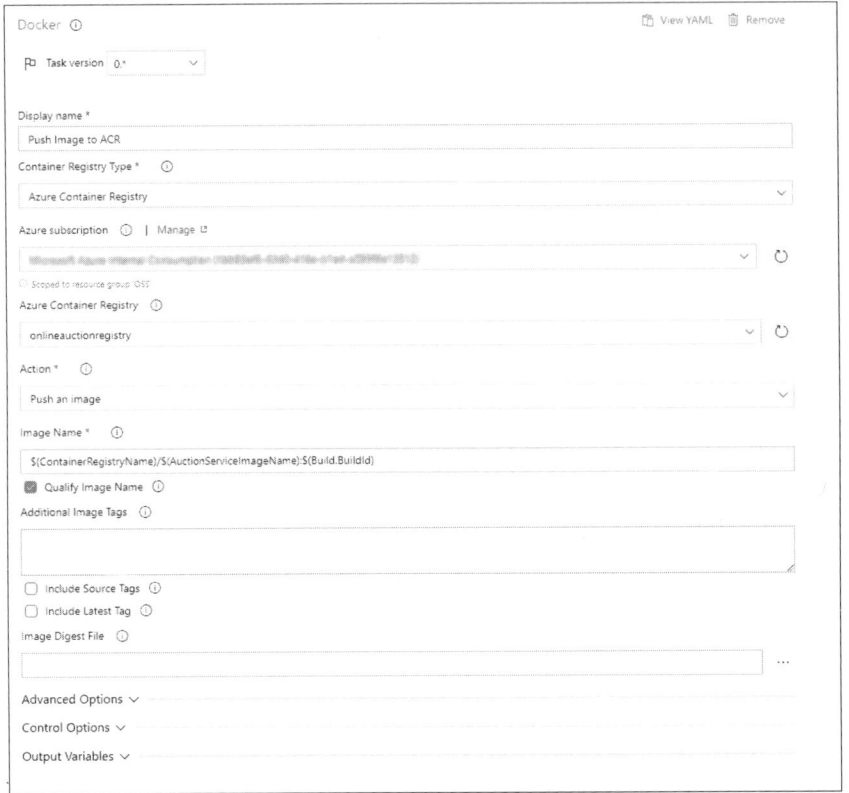

FIGURE 9-21 Push Image to ACR task

In this task, we are taking the image from the previous task and use a Docker push command to place the image in the container registry to deploy next.

Next, is the `kubectl apply` task, which is shown in Figure 9-22.

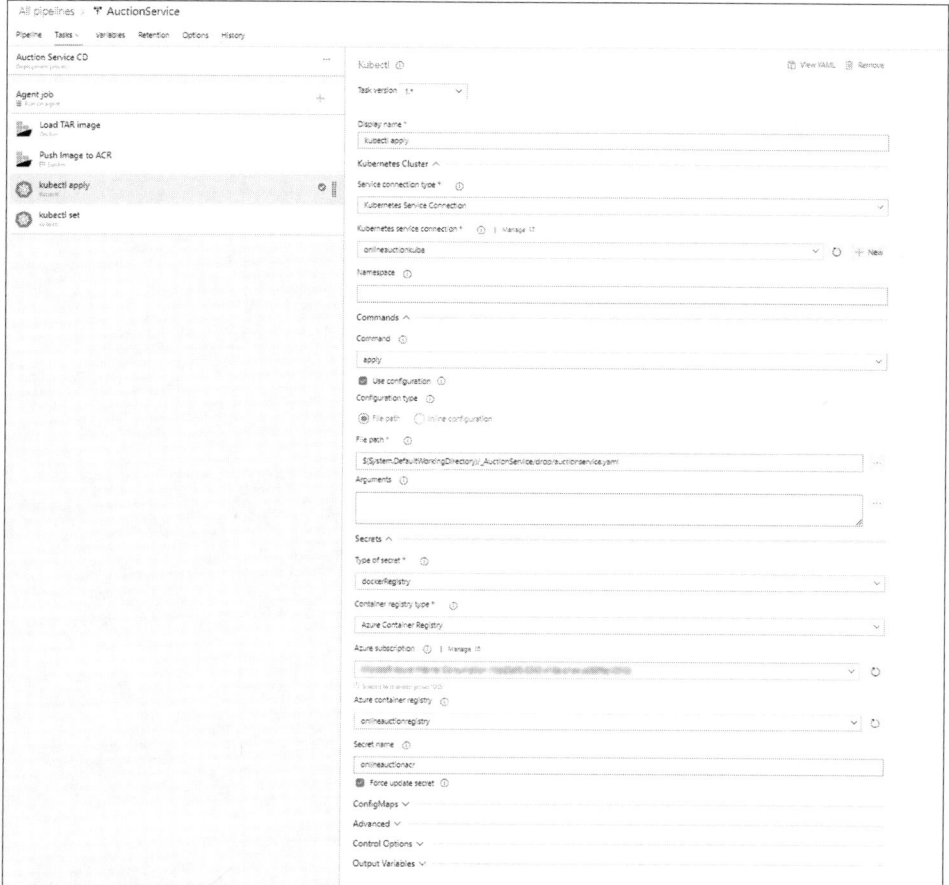

FIGURE 9-22   The `kubectl apply` task for the Auction Service deployment

This task kicks off the deployment of the container to Kubernetes with the Kubernetes service connection that we created in our settings. The Docker `secret` command helps manage the secrets needed to run the deployment. Here, we don't need to make an explicit function call because Azure DevOps already has a built-in capability to create a secret. The only important thing to note is that the secret should match the secret name included in the YAML file mentioned in the previous section and emphasized in Listing 9-2.

LISTING 9-2 Auction Service YAML file

```
apiVersion: apps/v1
kind: Deployment
metadata:
  labels:
    name: auctionservice
  name: auctionservice
spec:
  replicas: 3
  selector:
    matchLabels:
      name: auctionservice
  template:
    metadata:
      labels:
        name: auctionservice
    spec:
      containers:
        - image: onlineauctionregistry.azurecr.io/auctionservice
          name: auctionservice
          ports:
            - containerPort: 3000
        imagePullSecrets:
          - name: onlineauctionacr
---
apiVersion: v1
kind: Service
metadata:
  name: auctionservice
  labels:
    name: auctionservice
spec:
  ports:
    - port: 3000
      targetPort: 3000
      protocol: TCP
  type: LoadBalancer
  selector:
    name: auctionservice
```

Finally, run the kubectl set task, as shown in Figure 9-23.

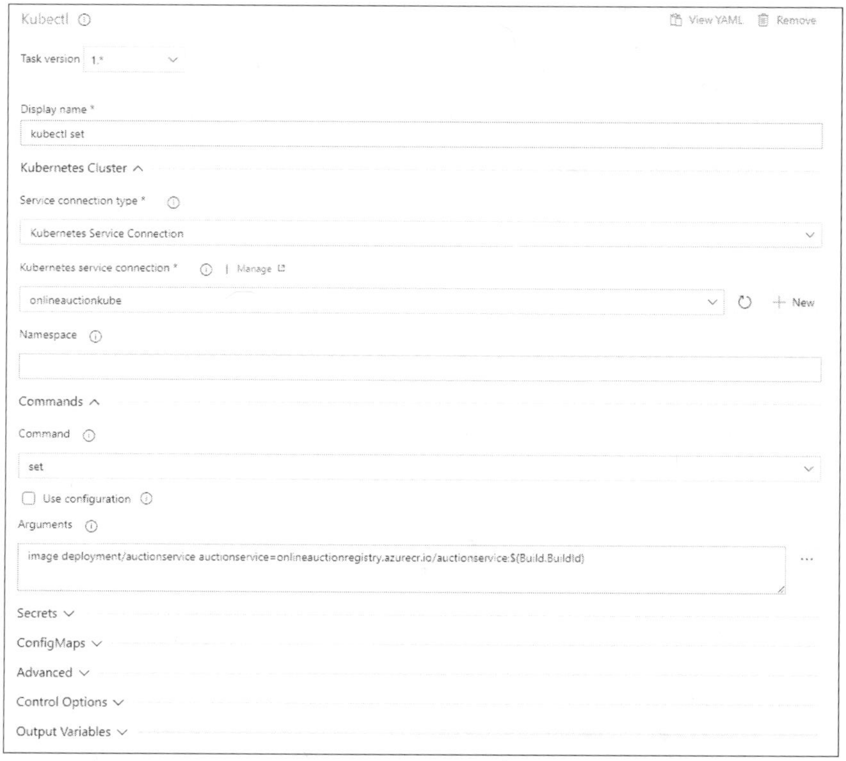

FIGURE 9-23 The kubectl set command

In this task, we use the Kubernetes Service Connection that we built earlier in this chapter and run the set command with the following argument:

```
image deployment/auctionservice auctionservice=onlineauctionregistry.azurecr.io/
auctionservice:$(Build.BuildId)
```

The image argument relates to what is being set, and then the deployment/auctionservice piece refers to the image property of the deployment kind; the auctionservice piece is the deployment name. We set the auctionservice containers to the reference image in the container registry. Finally, the $(Build.BuildId) variable is the predefined version that we are grabbing in Azure DevOps from our builds.

> **NOTE** Remember that kubectl is just one of the ways to run these commands. You can also leverage tools like Helm, which has become very popular recently.

Speaking of variables, we also need to address predefined variables for our build and release pipelines. When taking a look at the configurations that are set with the running of the pipeline as well as substituted values, we see a great mapping opportunity that will help us further simplify our pipelines. Take a look at the **Pipeline variables** in the build shown in Figure 9-24.

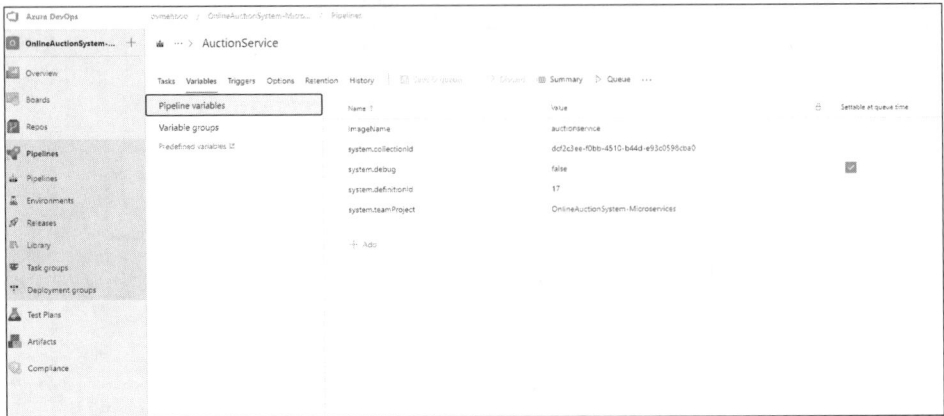

FIGURE 9-24 Pipeline variables for the auction service build

As you can see, several values have been initialized here for ease of use when building out the pipeline. Also, you can also do the same with variable groups. For example, you can input the name for a container registry that you regularly use.

Now we have a complete Azure DevOps pipeline that leverages CI/CD. If we make a change within the code base that we want to push out to our production auction service, we kick off these steps, and we will see those changes reflected when leveraging that service within the application.

As mentioned earlier, these examples demonstrate how to build these pipelines for your microservices. For the OAS, we used a similar set of pipelines for each of the other microservices, so be sure to check out the details within the code provided with this book.

A complete look at DevOps

So far, we looked at how to set up CI/CD using some real-life examples. As important as CI/CD is to the agility of an enterprise, it is important to have the right tooling to help us achieve these end-to-end scenarios. Azure DevOps has been that tool for us, but understanding how it can assist the overall process is important.

Following is an end-to-end summary of building the Azure DevOps CI/CD pipeline:

1. We start with an IDE or whatever developer tools (such as Visual Studio Code) you are using to build out your code base.

2. Next, you want to have individual code repositories to check your code into because you will likely have different microservices and each service gets its own repo. This can

be accomplished with tools such as GitHub or Azure DevOps, which could be cloned into local machines to pick up code changes committed to the code base and represented with version control.

3. Then we want this addition in the code base to automatically trigger our build using the examples demonstrated in this chapter, with the stable build ultimately moving forward to the Azure Container Registry to be pushed out to production. This can be seen in Figure 9-25.

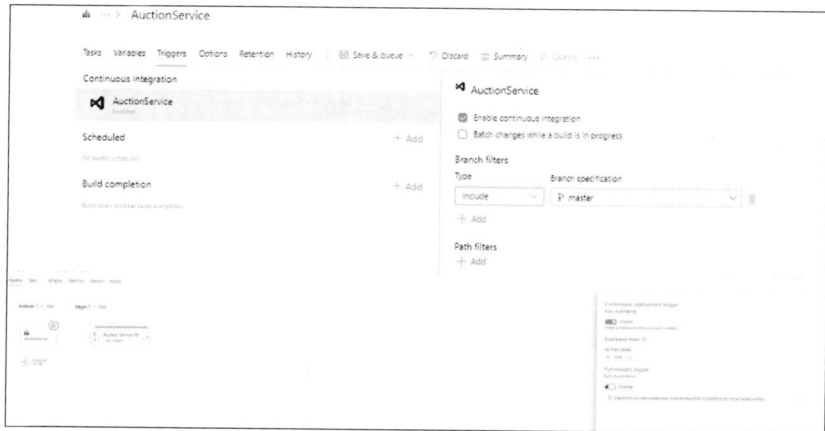

FIGURE 9-25 CI and CD triggers

4. These microservices are packaged as containers when they are pushed, which is done with Docker commands as seen in Chapter 5.

5. This push is performed with the deployment of the services via containers to the AKS cluster and to various pods inside. We have used the kubectl command `set image deployment` as a way to bind newer versions of images stored in ACR to AKS.

Now your users can access the services via apps and websites that consume these services by leveraging identity tools like Azure Active Directory B2C as a way to securely access the resources.

Summary

In this chapter, we accomplished the following:

- We achieved an understanding of continuous integration and deployment, as well as its importance.

- We looked at an example of CI/CD using IaC to illustrate the benefits, and we set up build and release pipelines.

- We learned how to create the build and release pipelines for the microservices for the OAS.

- We tied everything together with an understanding of Azure DevOps and the end-to-end deployment cycle.

There are great takeaways here with CI/CD and addressing some of the fundamental problems that enterprise workers see when using traditional monolithic architectures rather than microservices.

In Chapter 10, "Monitoring microservices," we will cover the fundamentals and best practices of monitoring. We will configure Azure Application Insights for the application and microservices and use Azure Monitor with other examples that will highlight the complete enterprise microservices architecture.

Monitoring microservices

After completing this chapter, you will:

- Understand the core concepts and patterns regarding monitoring and its importance in microservices

- Be able to configure Azure-oriented monitoring tools, such as Azure Application Insights and Azure Monitor to accomplish this in your own architectures as well

You now have all of the facets of your working architecture in place—congratulations! You have utilized the methods and theory presented in this book to help accomplish this, though the architecture is still technically incomplete without a monitoring framework. A monitoring framework will help us specifically address issues with applications, VMs, Kubernetes clusters, databases, and more.

In Chapter 1, "Introduction to microservices," we reviewed many of the pros and cons of microservices, and we mentioned monitoring frameworks as part of that list. When dealing with independent services built by independent teams, it is important to gather telemetry from both the individual and aggregate teams to understand where bottlenecks happen so you can refine the application. This helps reinforce the core ideas mentioned in Chapter 9, where you learned to build and deploy with ease and performance when do you learn of bottlenecks or performance issues, making you more reactive and agile.

In this chapter, we will cover the concepts and patterns regarding monitoring and then apply them to microservices. This will be done by walking through different configurations and technologies, such as Azure Application Insights, Azure Monitor, and more.

Monitoring concepts and patterns

There are several choices when it comes to application monitoring, but we found that certain technologies and frameworks that work particularly well for microservices, and you need to understand the native features available from the cloud provider you are using. Building modern, cloud-native applications is vital when transitioning from a monolith architecture to microservices, so you need to understand the monitoring capabilities that are incumbent to your cloud platform. One way to address this may be to have a cloud architect on your team who is knowledgeable about these capabilities.

A framework can cover several scenarios and cases. For example, log ingestion and collection engines are valuable for gathering log data from multiple sources and collating them into a singular pool.

Visualization tools are important as well, because it is hard to determine any trend with raw data alone. Visualization makes this easy; ultimately, integration with pre-created graphs and charts will help simplify and accelerate this process to provide faster value to your developers and testers.

For example, alerting systems use this gathered data and pattern matching to enable users to prioritize and react to certain events that could be considered abnormal or anomalies. This again will provide a reaction that ultimately makes the application better and thus a greater experience for your users.

Monitoring should be able to perform at least these basics: collect log information; create a methodology to analyze and visualize it; and generate performance metrics and details on your overall application and individual services. Each of these areas glosses over plenty of complexity, so we will walk through the theoretical aspects first before delving deeper into them.

Log information

Within your application, you will want to ensure that logs are generated for the many actions that could take place; you want to generate overall logs, as well as logs for the individual microservices. As such, there are two main focal points:

- Always code with logging in mind to help with troubleshooting.
- Make sure those logs can be aggregated somewhere for analysis when needed.

Many tools exist for collecting logs for retention purposes and troubleshooting analysis. These technologies can compile these lists with relevant details, such as the timestamp, operation, event initiation dependency, and more. There is great variety in the parameters of a log file; the important thing is to ensure that you are collecting and retaining them. To illustrate this, we will use Azure Log Analytics.

Azure Log Analytics

Azure Log Analytics is a resource that is related to Azure Monitor (We will discuss Azure Monitor later in this chapter). It is vital to use Azure Log Analytics to collect the log information from various parts of cloud-native applications. You can create a Log Analytics workspace in the Azure portal if you are leveraging Azure to create your microservices application.

> **TIP** Be sure to leverage Log Analytics as much as possible if you are creating any PaaS or IaaS resources in Azure because it is quite rich when combined with the Kusto Query Language (KQL).

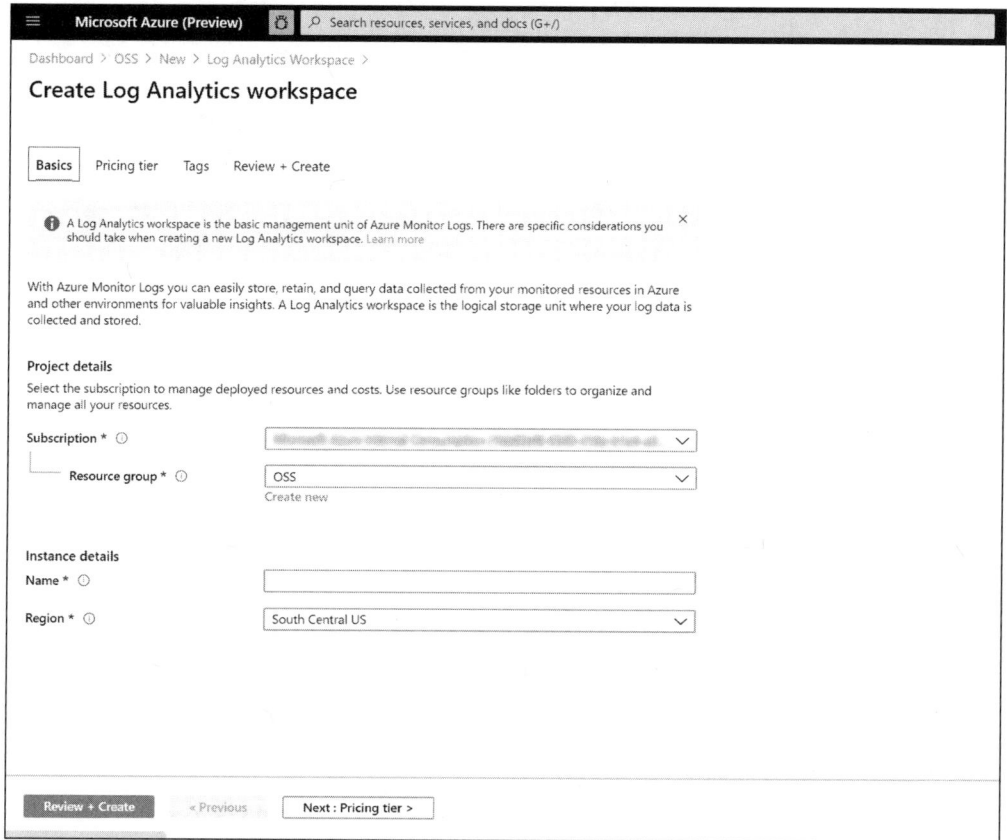

FIGURE 10-1 Creating a Log Analytics workspace in the Azure portal

In Figure 10-1, we see how to create a Log Analytics Workspace within the Azure portal, and it is quite similar to the creation details required for other Azure resources as well. You should also note that you can provision this workspace from tools like ARM (Azure Resource Manager) and IaC tools such as Terraform.

Figure 10-2 demonstrates an example of this workspace. In addition to creating this workspace, you also want to use any relevant connectors that can pipe these logs over to Log Analytics. These connectors, or agents, are important pieces for centralizing your logs and thus be able to leverage them.

> **NOTE** You will notice in some of the figures in this chapter that certain information is blacked out or replaced with an 'X'. This is to ensure that we do not reveal our entire subscription ID or Instrumentation Keys. We have done this for privacy and security reasons.

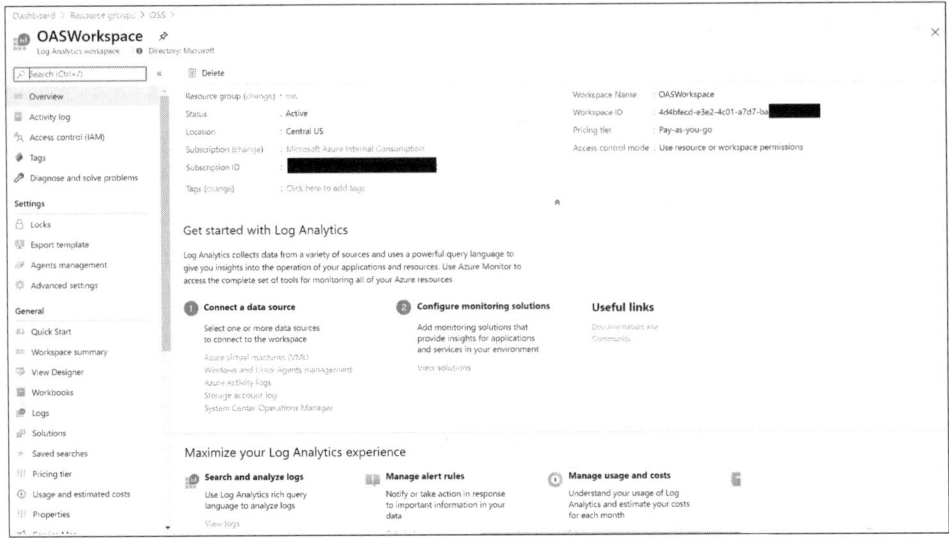

FIGURE 10-2 A Log Analytics Workspace named OASWorkspace

It is also important to note that your application logs could have their own particular schema when being transferred to the workspace. If your schema in the workspace is different from the one for your application, you will need to transform the schema to standardize it for the workspace. When this is done, you will be able to search and query for particular logs and information and further analyze them.

Retention is also important, and although there is a default of 90 days before logs are deleted, you can choose different versions to ensure that you have the retention you need for your organization.

Querying

Now that you have logs within your workspace or collection area, you can gather relevant data by querying. Log Analytics allows you to query logs using the Kusto Query Language (KQL). KQL helps you find the granular logs that will help you troubleshoot problems, and it can help you collect only the information necessary within the entire populace of logs with particular conditions. See the following query:

```
Event
| where EventLevelName == "Error"
```

Here, we are looking for particular events for which the `EventLevelName` parameter is equal to `Error`. Ultimately, KQL allows you to gather swaths of information with flexibility and ease. Depending on the tools you are using, there are many ways you can query for the information you want, so be sure to understand how the tool you are using works.

> **TIP** Holding training sessions on the querying language is useful for your developers, architects, and consultants, as well as your PMs, business leaders, and more. The gathered telemetry can be visualized tabularly or graphically with tools such as PowerBI.

Azure Monitor

Monitoring frameworks should be adopted for microservices. Many monitoring frameworks exist in different tools and frameworks, but for cloud-native applications, it can be useful to use the related feature set. Because we are discussing Azure as our primary cloud provider, we will discuss the relevant monitoring tool—Azure Monitor.

Azure Monitor can be thought of as a monitoring framework with many different components. A log aggregation pattern—a best practice for microservices—can serve as the primary telemetry and tracing pipeline that connects the rest of your Azure services, so there are many sources from which you can pull monitoring data. This data includes metrics, logs, health information, service events, and more.

We can work with these varieties of data from a single pane of glass because Azure Monitor is configured in the Azure portal, which means we have a consistent monitoring experience. Furthermore, we can even gather all the information from different data sources and visualize that data in custom-built dashboards that provide the relevant information that you might need regarding your application right away.

In Figure 10-3, we depict how Azure Monitor provides us a way to take information from containers, AKS, and operating systems to actionable results for developers and engineers.

From these sources, we can funnel information in the form of logs, metrics, or diagnostic logs. These could be captured through observations, audit logs, performance, events, and more. After pooling these, we can garner value in the form of insights, visualizations, and reactions. These can consist of features such as Container Insights, visualizations of workbooks, views, dashboards, and more. You can then use alerting and autoscaling systems to react to some significant events that can be garnered through the visualization, which informs engineers when making changes and improving the application for users.

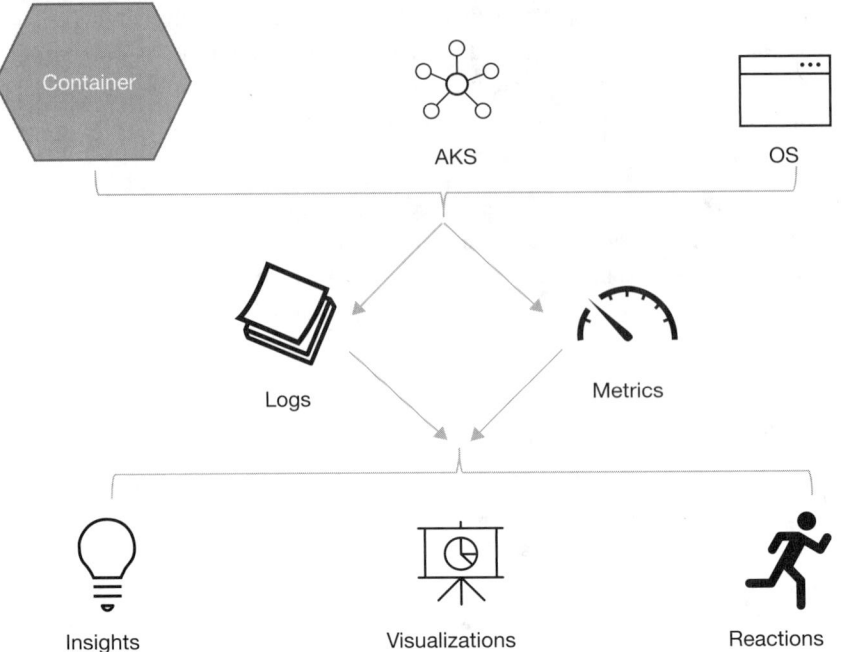

FIGURE 10-3 End-to-end functionality of Azure Monitor

In addition to such features, Azure Monitor is often improved and expanded. For example, the addition of Azure Cosmos DB allows you to pull information and place it in dashboards and visualizations. (We used Azure Cosmos DB with the OAS, which we will demonstrate later in this chapter.)

Azure Application Insights

Azure Application Insights is an Azure Monitor feature that can be deployed as an individual resource. It is focused on gathering live application performance telemetry to provide an instantaneous look at different facets of your application. Some of these insights include the following:

- Response rates for web pages
- Failures
- Exceptions
- Performance bottlenecks
- Sessions
- CPU, memory, and network usage
- Diagnostics
- Trace logs
- Customized metrics

Overall, Azure Application Insights is one of the most important tools at your disposal because of its rich feature set. We used Insights for our independent microservices because it coincides well with the concept of centralized logging and we could use it as a hub from which to collect data.

Also, this data collection enables us to calculate the performance metrics for all the services so we can react to any lapses with individual services and make changes to the build and deployment. This lifecycle becomes more seamless because of the independent nature of these services, and it enables you to react more quickly to deployment issues.

Take a look at Figure 10-4, which shows the **Investigate** blade in Azure Application Insights. Each of these options allows us to configure the types of insights as we mentioned above. For a breakdown of an application map and application resources, you can leverage the **Application map**. **Smart Detection** helps automatically identify any anomalies present in your application. You can then dive deeper using **Live Metrics, Search, Failures, Performance,** and more capabilities to gain exact insights into why your application might be not functioning optimally.

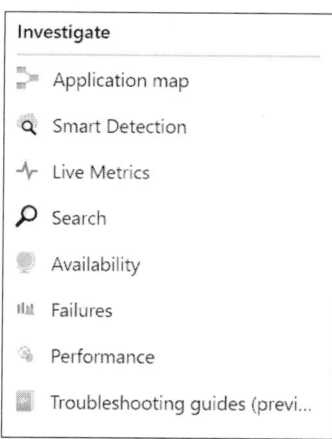

FIGURE 10-4   The Investigate pane in Azure Application Insights

The **Monitoring** section of Azure Application Insights is shown in Figure 10-5. Here, we can be more granular and query for individual logs as part of Azure Monitor. Thus, you could use specific Kusto queries in the **Logs** selection. You can also set and configure alerts and specific triggers by leveraging **Alerts**, which is a staple of Azure Monitor. This helps leverage the data-rich ecosystem to find customized and specific application events and react to them automatically through the use of workbooks or even manually through a flow that is kicked off to notify testers and developers.

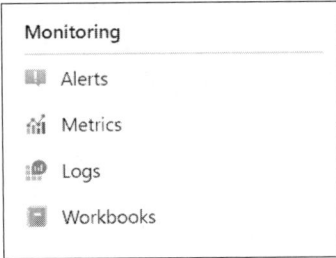

FIGURE 10-5 Monitoring pane in Azure Application Insights

In Figure 10-6, the **Usage** section is shown. Here, we can see the usage details in conjunction with retention information, and we can find specific telemetry on particular sessions and events.

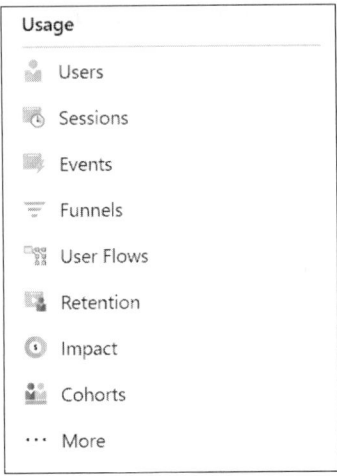

FIGURE 10-6 Usage pane in Azure Application Insights

Figure 10-7 shows the **Configure** options in Azure Application Insights. This is where you can get the specific information on the details behind the setup of your application resources, such as the **Usage and estimated costs**, **API Access**, **Properties**, and more.

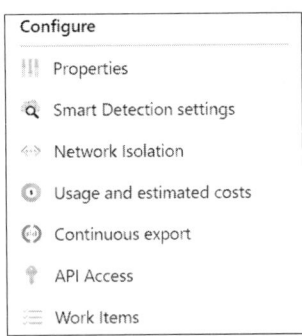

FIGURE 10-7 Configure pane in Azure Application Insights

We will go deeper into the configuration of Azure Application Insights in the following sections as we demonstrate what we did with the OAS application.

Monitoring framework and best practices

Think back to our original architecture diagram from Chapter 3, which is shown again in Figure 10-8. As seen here, we leveraged the monitoring framework from Azure Monitor. Now that you have more understanding of the technologies underlying monitoring with Azure, let's put it into context of our OAS to show you exactly how to build out everything within a microservices architecture.

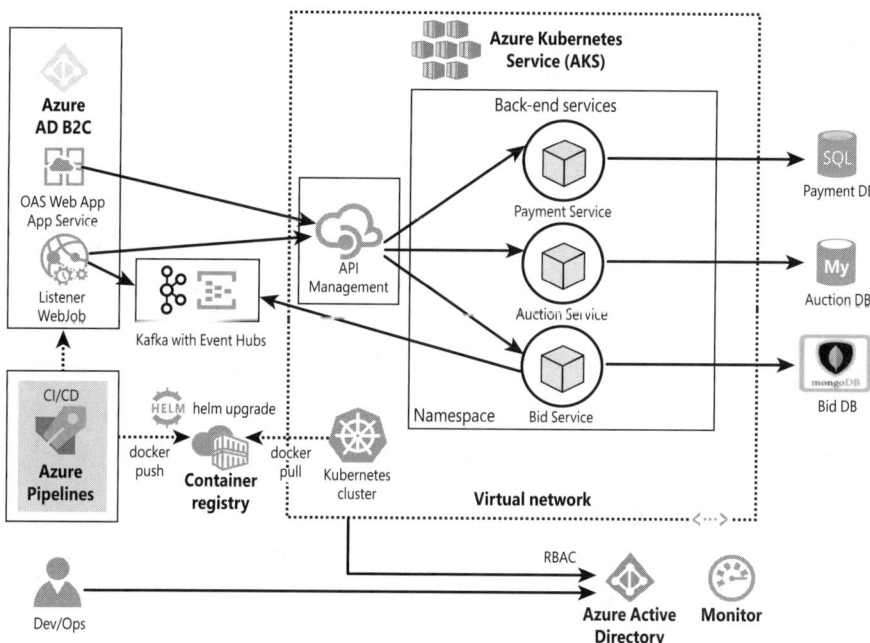

FIGURE 10-8 End-to-end microservices architecture for the OAS application

In our case study, we leveraged Azure, which means our application metrics and performance are recorded by tools such as Azure Monitor, Log Analytics, Azure Application Insights, and much more. Whatever your platform might be, the key to addressing a potential challenge is to be informed and intelligent about the different options available to you—be they cloud-native, third-party tools, or open-source tools—to make good decisions for your monitoring frameworks.

Note that we are leveraging the Log Aggregation framework, which focuses on centralizing logging and aggregates logs from each service instance. In our case, we have many components: each of our line-of-business microservices, our Angular App for our front-end, and more. We will go through our setup to help you understand how you could approach this as well.

> **TIP** As a general rule, the complexity of your monitoring increases as your services increase. A framework such as Log Aggregation will help you address the difficulties that come from scaling out infrastructure from monitoring.

Azure Application Insights configuration

A good rule to follow when it comes to monitoring is to focus on logging and to ensure that you centralize it. With a microservices application, you may run into issues related to network transient failures or latency problems, so to combat these issues we have leveraged Azure Application Insights for each of our components. Although you could have an Azure Application Insights instance for each component that you would include in the application, this might not always be the best way to approach this issue. Furthermore, it's vital to ensure that these instances are being funneled in a singular store, such as Azure Log Analytics. Multiple instances make it harder to troubleshoot because of management issues, so consolidation is a great tactic to use here.

We specifically leveraged Azure Application Insights and integrated it with the rest of our Azure components and our microservices. Take a look at Figure 10-9, which shows one of our Azure Application Insights instances named **MicroservicesInsights;** this instance gathers telemetry on our three core microservices (auction, bid, and payment).

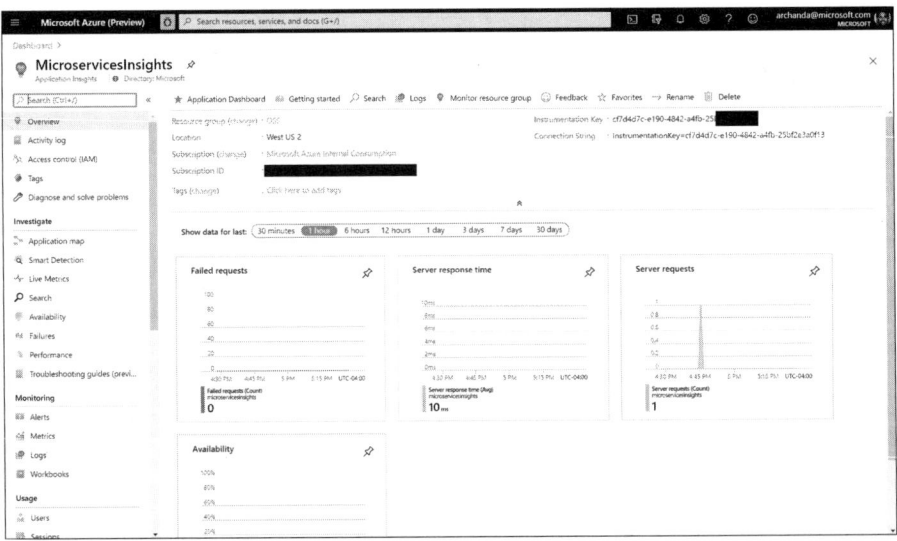

FIGURE 10-9 Azure Application Insights instance named `MicroservicesInsights`

We leveraged Azure Application Insights and integrated it with our core microservices. As you know, we have three core microservices: auction, bid, and payment. Each of these are made with different technologies, so we wanted to demonstrate how to configure Azure Application Insights given this variety in technologies.

Auction service

As you remember, the auction service was built using Node.js, so we will go through the process of how to use the Azure Application Insights SDK in conjunction with Node.js to get telemetry going to this particular Azure Application Insights instance.

> **NOTE** The actual creation of an Azure Application Insights instance is quite straightforward and doesn't need many components. After you create the instance, take note of the **InstrumentationKey**, which is important for many of the configuration steps we will undertake.

For Node.js, we want to focus on the dependencies for the configuration—these have to be run on the root path of the auction service where the `package.json` file resides. Use the following command to install Azure Application Insights.

```
npm install applicationinsights -save
```

Within your dependencies, you should have the following construct with some variability on the version of Azure Application Insights that you can use:

```
"dependencies": {
    "applicationinsights": "^1.6.0",
    …
}
```

This dependency will declare that you are using Azure Application Insights within your Node.js application. You must finish the configuration by including the following code snippet in your `index.js` file:

```
let appInsights = require("applicationinsights");
appInsights.setup("cf7d4d7c-e190-4842-a4fb-25bf2e3xxxxx")
.setAutoDependencyCorrelation(true)
.setAutoCollectRequests(true)
.setAutoCollectPerformance(true)
.setAutoCollectExceptions(true)
.setAutoCollectDependencies(true)
.setAutoCollectConsole(true)
.setUseDiskRetryCaching(true);

const key= appInsights.defaultClient.context.keys.cloudRole;
appInsights.defaultClient.context.tags[key]='AuctionService';
appInsights.start();

let client = appInsights.defaultClient;
```

In this code, the first line contains the `require` method, which loads the `applicationinsights` module. In the setup method, we use the InstrumentationKey as the ID representing the Azure Application Insights instance to which we want to funnel our metrics. Remember, we are funneling the telemetry for all three microservices to a single instance, so we will be seeing more of this key. Following this, we have a set of commands that set up the kinds of telemetry that we are collecting, such as `Performance`, `Requests`, `Exceptions`, and so on. The last code line tags the telemetry, associating it as `AuctionService` telemetry and then initializes the client.

Now let's take a look at how the configuration is used in the rest of the code. For that, we need to navigate further down on the `index.js` page. In the following code, we have several functions that dictate the functionality of the auction service. One such function updates the auction data that is sent through to the bid topic.

```
//Update auction data sent through Kafka consumer for topic: bidtopic

router.put('/updateAuctionForBid', (request,response)=>{

  client.trackTrace("Updating auction for bid");

  console.log("bid id is "+ request.body.bidid) ;

  console.log("bidPrice is "+ request.body.bidAmount);

  console.log("auctionId is "+ request.body.auctionId);

  console.log("userId is "+ request.body.userId);

  pool.query("UPDATE auction set bidId='"+request.body.bidid+"', bidPrice="+request.
  body.bidAmount+", bidUser='"+request.body.userId+"' where idAuction="+ request.body.
  auctionId, (error, result) => {

    if (error) throw error;

    response.send("updated");

  });

});
```

Here, along with the other functions, we have leveraged the `trackTrace` method of the Azure Application Insights client object as part of gathering telemetry at different phases of the application. Although we used the `trackTrace` method here, we also can use functions such as `trackDependency`, `trackEvent`, and `trackRequest`.

> **NEED MORE INFO?**   For the full breakdown of the possible methods that you can use with the `TelemetryClient` object, see the Microsoft documentation at *https://docs. microsoft.com/en-us/dotnet/api/microsoft.applicationinsights.telemetryclient?view=azure-dotnet*.

Bid service

Now let's take a look at the bid service configuration, which was built using Java. To get started, we need to view it from the perspective of dependencies. First, let's take a look at the `pom.xml` file and add this particular dependency within the `dependency` section:

```
<dependency>
          <groupId>com.microsoft.azure</groupId>
          <artifactId>applicationinsights-spring-boot-starter</artifactId>
          <version>1.1.1</version>
</dependency>
```

Then we move to the `BidController.java` file to import the following package:

```
import com.microsoft.applicationinsights.TelemetryClient;
```

This allows us to utilize the dependency created earlier within the code. Now, within the rest of the code, we initialize the `TelemetryClient`, as we did earlier.

```
@Autowired
    TelemetryClient telemetryClient;
public ArrayList GetBidByAuctionID(String auctionID)
        {
            ArrayList lst=new ArrayList();
            telemetryClient.trackTrace("Getting Bid by Auction ID");

...
```

After leveraging this, we can now use functions such as `trackTrace`. For more detail, see the complete `BidController.java` file in the code repository associated with this title in the Microsoft Press website.

Payment Service

The payment service was created using .NET Core, so we need to leverage the NuGet package manager to install the `Microsoft.ApplicationInsights.AspNetCore` package, which allows us to leverage Azure Application Insights within .NET Core. Also, we created a custom `logger` class here to gather and funnel logs to a multitude of services that engineers can consume and use. Currently though, we only have set up Azure Application Insights here as the source.

Now let's navigate to the customer logger class we created in `logger.cs`. At the top of this class file, we used the package that we installed by using this code:

```
using Microsoft.ApplicationInsights;
using Microsoft.ApplicationInsights.Extensibility;
```

We can now take a look at the code from the rest of the file, as shown in Listing 10-1:

LISTING 10-1 Logger Class

```
public class Logger
    {
        private static Logger _instance;
        TelemetryConfiguration configuration = TelemetryConfiguration.CreateDefault();

        private Logger()
        {
            configuration.InstrumentationKey = "cf7d4d7c-e190-4842-a4fb-25bf2e3xxxxx";
        }
```

```
    public static Logger Instance
    {
        get
        {
            if (_instance == null)
            {
                _instance =  new Logger();
            }
            return _instance;
        }
    }

    public void LogMessage(string message)
    {
        var telemetryClient = new TelemetryClient(configuration);
        telemetryClient.TrackTrace(message);
        Console.WriteLine(message);
    }
}
```

This is part of the code for the custom logger. As you can see, we created a Logger constructor and instance, and then we created the configuration. The main part of this configuration is to ensure that the InstrumentationKey used earlier is consistent with this one. Then at the bottom, we have a LogMessage method that takes a string message as a parameter and sends custom trace telemetry to Azure Application Insights. The Logger class is a singleton class, so only this instance will be shared across all the classes. In order to use this class, the following example shows how we are using the respective logger method:

```
// GET: api/AuctionPayments
    [HttpGet]
    public async Task<ActionResult<IEnumerable<AuctionPayment>>> GetAuctionPayment()
    {
        Logger.Instance.LogMessage("Getting auction payment");
        return await _context.AuctionPayment.ToListAsync();
    }
```

In this example, we have the logger instance line from AuctionPaymentsController.cs where we use the logger object to log this message: Getting auction payment. We use this throughout this controller file, which you can find in our code repository.

Angular front-end

Now we want to turn our attention to a new Azure Application Insights instance that was created for our OAS front-end using Angular. Figure 10-10 shows an Azure Application Insights instance named onlineauctionwebapp, which gathers telemetry from our front-end component that was made with Angular.

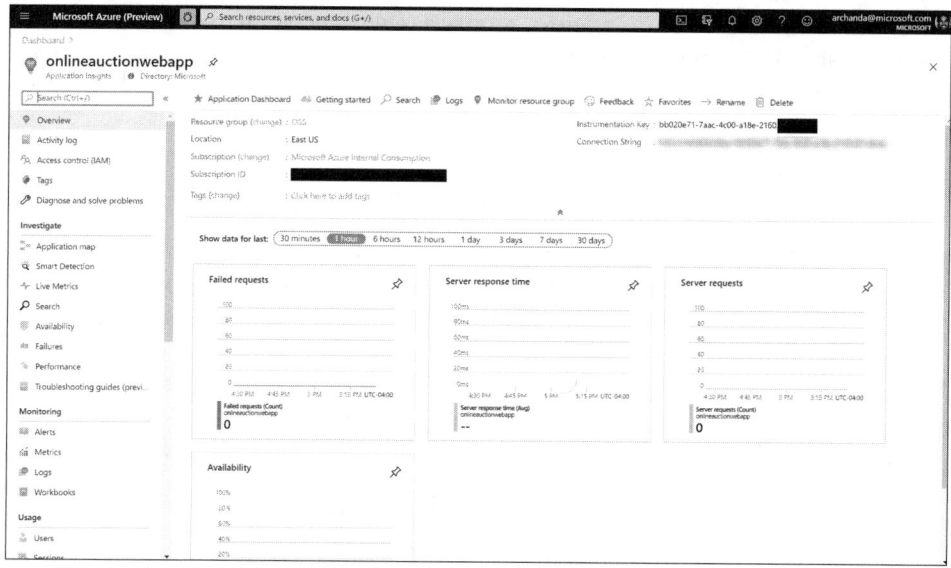

FIGURE 10-10 An Azure Application Insights instance named onlineauctionwebapp

To configure this instance of Azure Application Insights, we want to add a dependency to the package.json file for the front-end code. Add this line to the list of dependencies:

```
"@microsoft/applicationinsights-web": "^2.5.3",
```

Then you want to create a TypeScript file—we named ours app-insights.service.ts. This enables us to put in code oriented for both Angular and Azure Application Insights for the constructor, instance, and more. See the following code:

```
import { Injectable } from '@angular/core';
import {ApplicationInsights} from '@microsoft/applicationinsights-web';
import { environment } from './environments/environment.prod';
@Injectable({
  providedIn: 'root'
})
export class AppInsightsService {

  instance: ApplicationInsights;
  constructor() {
      this.instance = new ApplicationInsights({config: [
          instrumentationKey: environment.appInsights.instrumentationKey
      ]}) ;
      this.instance.loadAppInsights();
      this.instance.trackPageView();


  }
}
```

Once again, we have a `constructor` here in which we reference the `instrumentationKey`, in which we will also store in the environments file we named `environment.prod.ts`. This time, make sure to replace the `instrumentationkey` with your new instance:

```
appInsights:{
    instrumentationKey:"bb020e71-7aac-4c00-a18e-xxxxxxxxxxxx"
}
```

Now let's move over to seeing this configuration in use by navigating to our file called `auction.component.ts`. In this file, we have the `CreateAuction` method to which we can then add the following:

```
this.appInsights.instance.trackEvent({name: 'CreatedAuction'});
```

Now we can leverage the library to use functions like `trackEvent` to garner the telemetry we need. As always, you can reference the full code in the `auction.component.ts` file by viewing the code repository.

Kafka Hosted Service and Azure Application Insights configuration

Now we get to our final Azure Application Insights instance we made for our OAS, which was built for the Kafka Hosted Service using .NET Core (see Figure 10-11).

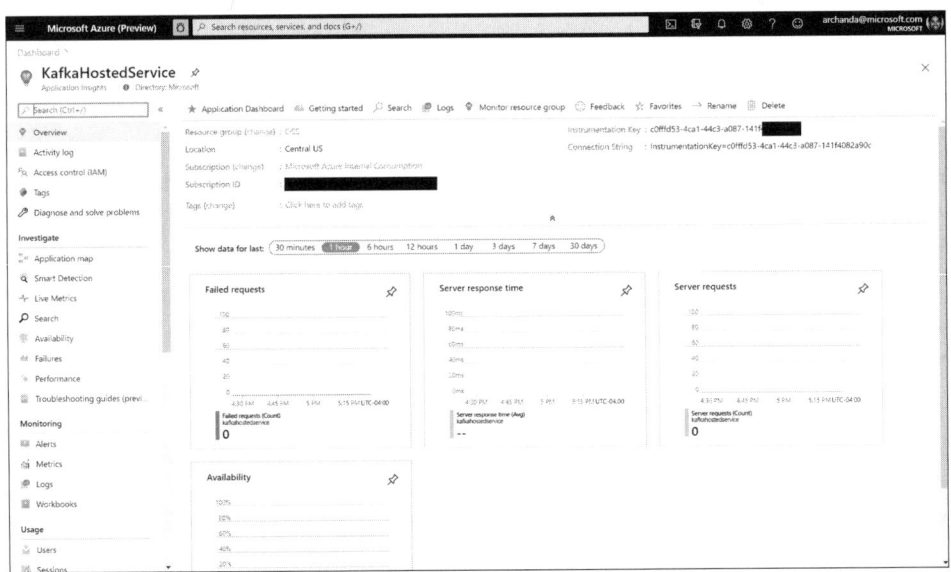

FIGURE 10-11   Azure Application Insights instance named `KafkaHostedService`

Because this service is built using .NET Core, we will use the same approach we used earlier in this chapter with the payment service. This means that you want to use NuGet Package Manager to acquire the `Microsoft.ApplicationInsights.AspNetCore` package and then use the related packages within the code to create a logger class and subsequently use those functions to log messages and telemetry from this service. It is a very similar pattern to the payment service, so we recommend that you take a look at the relevant code in the repository for more specific information.

Container Insights

In many microservices architectures, you will need to use containers and container orchestration. For the infrastructure used in microservices, it is really useful to gather infrastructure telemetry to understand constraints on your application such as scalability and implement autoscaling solutions such as increasing the number of pods when needed.

This information can be gathered and visualized with the other insights that you will receive, and juxtaposing them together would be useful in revealing correlating events. In the next section, we will focus on dashboards as the final piece of your monitoring framework. We collected the telemetry using a Log Analytics Workspace and created a solution based on it, as shown in Figure 10-12.

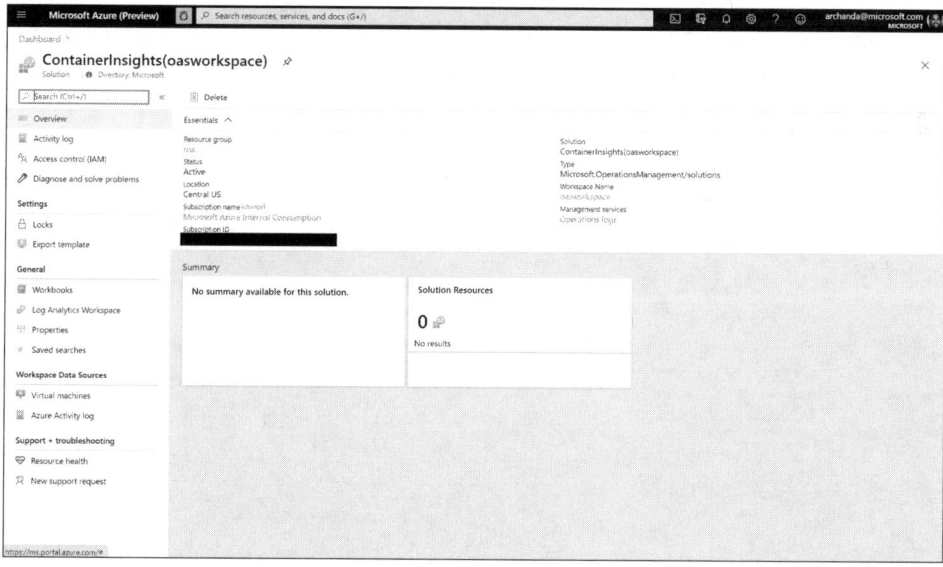

FIGURE 10-12 A Log Analytics Workspace to gather information on container workloads in the OAS

Another point to note is the importance of using container orchestration in an microservices architecture. You learned in Chapter 8, "Set up Azure API Gateway," about the importance of patterns and best practices in this area, but we also need to understand how to monitor them.

To this end, we also leveraged Azure Monitor for monitoring containers, and there are several ways to implement this. When creating your AKS cluster and infrastructure using the Azure portal, you can enable container monitoring as part of the workflow, as shown in Figure 10-13.

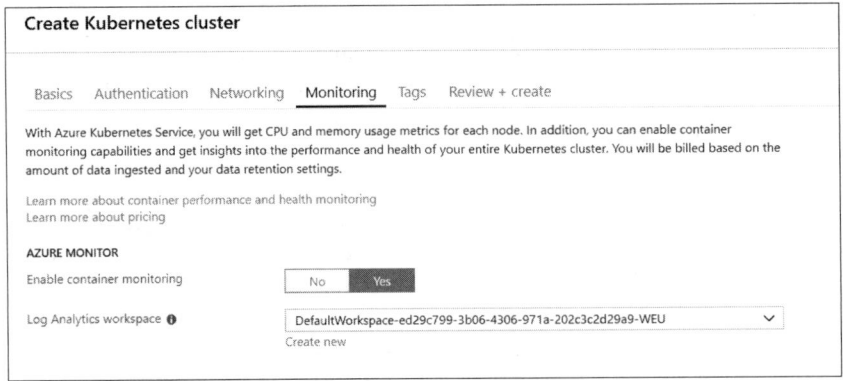

FIGURE 10-13 Enabling container monitoring as part of the workflow

If you were building this instead through scripting, you can use the shell in Azure with the Az module to run the following command to enable monitoring on an existing AKS Cluster:

```
az aks enable-addons -a monitoring -n oasAKSCluster -g OSSRG
```

Finally, you can enable monitoring of one or more AKS clusters already deployed from the multi-cluster page in Azure Monitor, as shown in Figure 10-14.

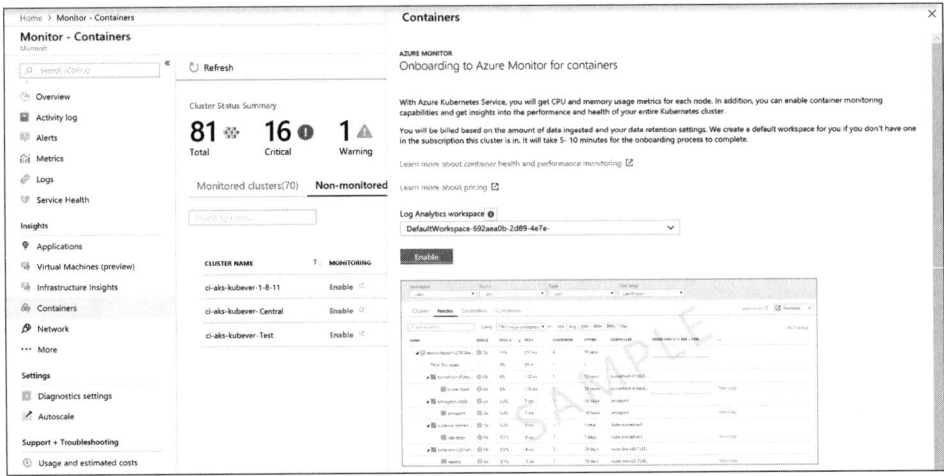

FIGURE 10-14 Setting up monitoring for a container

Dashboards

The last piece of our monitoring framework for the OAS application is to use dashboards to visualize all the information you have been collecting. Azure provides a great way to create dashboards that help you customize exactly what you want to see.

We specifically made two customized dashboards based on the `KafkaHostedService` and the front-end web app. You can see the two dashboards in Figures 10-15 and 10-16.

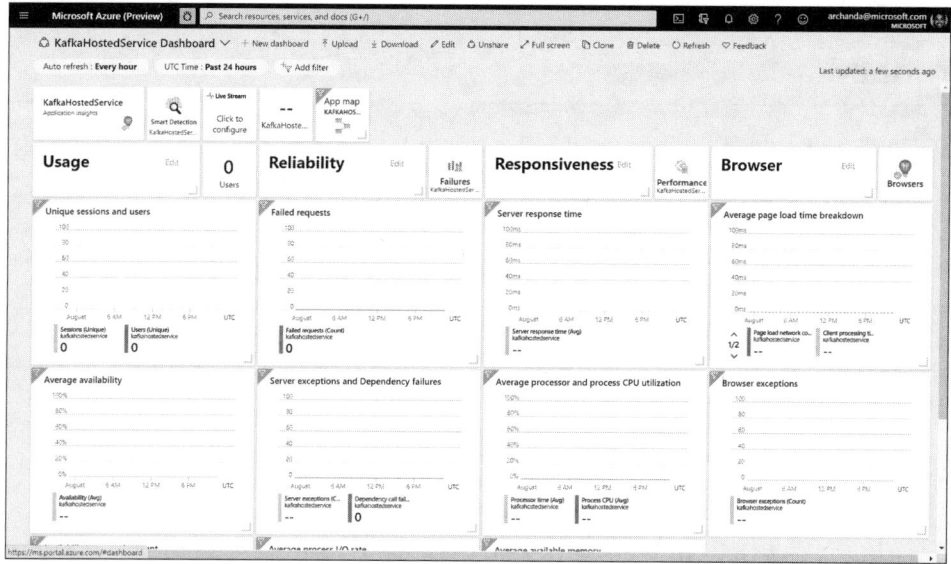

FIGURE 10-15 KafkaHostedService dashboard created for the OAS

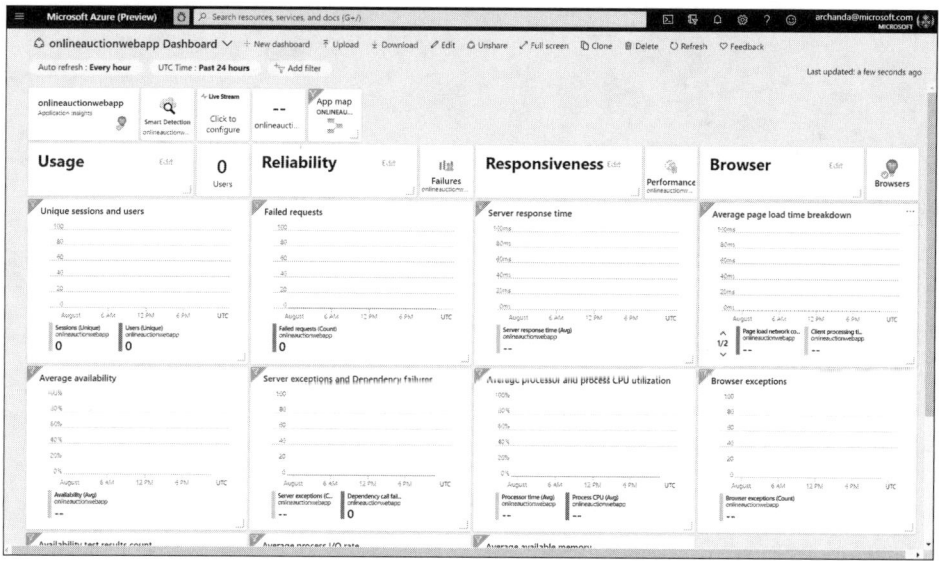

FIGURE 10-16 onlineauctionwebapp dashboard created for the OAS

With both dashboards we focused on four main tenets, which are usage, reliability, responsiveness, and browser. We checked for unique sessions and users over time along with the availability over time to determine usage metrics. For reliability, we checked for failed requests over time and exceptions/failures over time. Accordingly, we built out the graphs that we felt would be best to determine these tenets and would help us improve the application in each of these areas.

Summary

In this chapter, you:

- Learned about monitoring and its importance in microservices
- Learned how to configure Azure-oriented monitoring tools such as Azure Application Insights and Azure Monitor for the OAS to better build out your monitoring frameworks

Monitoring is a key feature for many applications, and getting insight into how to use it within microservices is important in your journey to a complete microservices architecture. Using these concepts and walkthroughs should provide the guidance needed to do it yourself.

Index

A

ACID (atomicity, consistency, integrity, durability), 50
ACR (Azure Container Registry)
  provisioning, 129
  pushing services to, 129–130
ADO (Azure DevOps), 49
aggregator pattern, 53–54
agile, 13
AKS (Azure Kubernetes Services), 1, 48–49
  clusters, 42–43
  configuring with APIM (Azure API Management), 218–220
  creating a deployment object for OAS microservices, 131–135
  creating a service object for OAS microservices, 135–137
  creating a YAML file, 131–133
  deploying images to, 125
  deploying services to, 130–131
  ingress controller-accessible services, 220
  Internet-accessible services, 218–219
  intranet-accessible services, 219
  provisioning, 127–128
  using with APIM, 214–215
    ClusterIP configuration, 216–217
    External Name configuration, 217–218

LoadBalancer configuration, 216
NodePort configuration, 215
AMPQ (Advanced Messaging Queuing Protocol), 146
Angular, 43, 89
  components, 94
  creating front-end applications, 89–90
  environment files, configuring, 101–102
  modules, 94
  onlineauctionwebapp, 262–264
  project structure, 90–92
  routing, 95
  services, 95
  templates, 95
anti-patterns, 29, 33. *See also* patterns
  Direct HTTP call, 52–53
  establishing tight dependencies between code artifacts, 34
  unnecessary fine graining of services to deeper subdomains, 33–337
  using monolith or a shared database with microservices, 33
Apache Kafka, 153–154
  adding Kafka support in the Java application, 159–161
  BidHostedService class, 164–165

building the HTTP client, 171–172
establishing pub/sub communication, 157
Event Hubs, 155–156
infrastructure setup, 157–159
KafkaHosttedService, 264–265
listener service with .NET Core hosted service, 161–164, 165–166
resiliency, 165–171
APIs, 47–48, 82, 138
  auction service, 66–68
  configuring with APIM (Azure API Management), 204–210
  creating a bid service using JavaSpring Boot framework, 72–74
  gateways, need for, 201–202
  methods used in bid service, 74
application development, DDD (domain-driven design), 27
  bounded context, 28, 29–30
  core domains, 30
  and decomposition, 31–32
  supporting domains, 30–31
  ubiquitous language, 27
Application Insights, 254–257
  auction service configuration, 259–260
  bid service configuration, 260–261

B

C

Hear about it first.

Since 1984, Microsoft Press has helped IT professionals, developers, and home office users advance their technical skills and knowledge with books and learning resources.

Sign up today to deliver exclusive offers directly to your inbox.

- New products and announcements

- Free sample chapters

- Special promotions and discounts

- ... and more!

MicrosoftPressStore.com/newsletters

 Pearson

Plug into learning at

MicrosoftPressStore.com

The Microsoft Press Store by Pearson offers:

- Free U.S. shipping

- Buy an eBook, get three formats – Includes PDF, EPUB, and MOBI to use with your computer, tablet, and mobile devices

- Print & eBook Best Value Packs

- eBook Deal of the Week – Save up to 50% on featured title

- Newsletter – Be the first to hear about new releases, announcements, special offers, and more

- Register your book – Find companion files, errata, and product updates, plus receive a special coupon[*] to save on your next purchase

 Pearson